Electrostatic Discharge Protection for Electronics

Electrostatic Discharge Protection for Electronics

Neil Sclater

TAB Professional and Reference Books

Division of TAB BOOKS
Blue Ridge Summit, PA

FIRST EDITION
FIRST PRINTING

© 1990 by TPR Books, an imprint of **TAB BOOKS**
TAB BOOKS is a division of McGraw-Hill, Inc.

The TPR logo, consisting of the letters "TPR" within a large "T," is a registered trademark of TAB BOOKS.

Printed in the United States of America. All rights reserved. The publisher takes no responsibility for the use of any of the materials or methods described in this use of any of the materials or methods described in this book, nor for the products thereof.

Library of Congress Cataloging-in-Publication Data

Sclater, Neil.
    Electrostatic discharge protection for electronics / by Neil Sclater.
      p.   cm.
    Includes bibliographical references.
    ISBN 0-8306-8329-1
    1. Electronic apparatus and appliances—Protection. 2. Electric discharges. 3. Electrostatics. I. Title.
TK7870.S375   1990
621.381—dc20   90-35298
                                                  CIP

TAB BOOKS offers software for sale. For information and a catalog, please contact TAB Software Department, Blue Ridge Summit, PA 17294-0850.

Questions regarding the content of this book should be addressed to:

Reader Inquiry Branch
TAB BOOKS
Blue Ridge Summit, PA 17294-0850

Vice President and Editorial Director: Larry Hager
Technical Editor: Andrew Yoder
Production: Katherine Brown
Series Design: Jaclyn J. Boone

Contents

Acknowledgments viii
Introduction ix

1 Static Electricity: A Growing Menace   1
What Is Electrostatic Discharge? 5
Electrostatic Discharge and Electronic Devices 7
Electrostatic Discharge as a Universal Problem 8
A Short History of Static Electricity 10
The ESD Environment 11

2 Principles of Electrostatics   15
Electron Theory and Atomic Structure 16
Electrostatic Attraction and Repulsion 17
Triboelectric Charging 20
Conductors and Insulators 24
Charge Dissipation 26
Charging by Induction 29
Electric Fields 30
Electrostatic Measurements: The Electroscope 34
Electrical Potential 40
Equipotential Surfaces 41
Surface Charge Density 43
Capacitance 44
Electrostatic Generators 50
Static Charge Dissipation 55

3 Electrostatic Discharge Damage Mechanisms   56
Susceptible Electronic Devices   57
Semiconductor Device Susceptibility Classes   57
Hard Semiconductor Failures   61
Electron Microscope Analysis   64
Soft Semiconductor Failures   65

4 Semiconductor Device Board and System Protection   68
Conditioning the Environment   68
Protection for Integrated Circuits   69
Protection for Semiconductors at Board Level   72
System Immunity to ESD   75

5 Protective Materials and Packaging   77
Materials Classification   78
Protective Bags   83
Protective Materials and Packaging   84

6 Protective Workplace Environment   93
Humidity Control   93
Protective Work Surfaces   95
Air Ionizers   104
Nuclear Ionizers   108
Tools and Production Equipment   110
Topical Antistats   111
ESD Labels and Signs   113

7 Personal Protection against ESD   114
Grounded Wrist Straps   114
Foot Grounding Straps   120
ESD-Protective Clothing   121
Gloves and Finger Cots   124
Field Service Kits   124

8 Electrostatic Test Equipment   **126**
   Portable Survey and Audit Instruments 127
   Theory of Electrostatic Fieldmeters and Voltmeters 127
   Electrostatic Monitors 140
   Surface/Volume Resistivity Probe 142
   Static Decay Meter 143
   Humidity Test Chamber 145
   Shielded Bag Test System 146
   Ground Strap Testers 147

9 Electrostatic Discharge Simulation Equipment   **149**
   ESD Standards 152
   Commercial ESD Simulators 153

10 The Complete ESD Control Program   **159**
   Preparation and Monitoring of an ESD Control Program 160
   Preparation for Work at an ESD-Controlled Workstation 161
   General Handling Procedures and Requirements 166
   General Guidelines for Handling ESD-Sensitive Devices 167
   General Guidelines for Handling ESD-Sensitive Circuit Boards 169
   Receiving Inspection Procedures for ESD-Sensitive Items 170
   Field Testing and Inspection 171
   Audit Provisions for ESD Control 172
   Personnel Training and Certification 173
   Product Design 174
   Customer Responsibilities 175

Appendices

  A Standards   **176**

  B Directory of Suppliers   **178**

  References   **199**

  Glossary   **201**

  Index   **221**

Acknowledgments

Protecting Electronic Equipment from Electrostatic Discharge by Edward A. Lacy, published by TAB BOOKS in 1984, was a valuable resource used in the preparation of this book.

Military Handbook DOD-HDBK-263, *Electrostatic Discharge Control Handbook for Protection of Electrical and Electronic Parts, Assemblies and Equipment*; Military Standard 883C, Notice 7, *Test Methods and Procedures: Method 3015.6 Electrostatic Discharge Sensitivity Classification*; Electronics Industries Association EIA-541, *Packaging Material Standards for ESD Sensitive Items*; and International Electrotechnical Commission, IEC Publication 801-2, *Electromagnetic Compatibility for Industrial Process Measurement and Control Equipment, Part 2: Electrostatic Discharge Requirements* are widely quoted and used as an information source for this book.

The following manufacturers and organizations graciously supplied information: Chapman Corporation; Charleswater Products, Inc.; Electro-Tech Systems, Inc.; KeyTek Corporation; Monroe Electronics, Inc.; NRD Inc.; The Simco Company, Inc.; The Static and Electromagnetic Control Division of 3M Corporation; and Static, Inc., A Sealed Air Company.

Introduction

Electrostatic discharge (ESD) is a leading cause of electrical overstress (EOS) for electronic components and systems. It exacts its toll in many ways, from destroying semiconductor devices to latching-up computer systems. It is also responsible for causing latent damage to integrated circuits (ICs) that might escape detection during routine inspection. Later, the damage causes failure in the host equipment, necessitating expensive and time-consuming field repairs.

Replacing an ESD-sensitive component that has failed *on duty* is far more costly than replacing one that has been destroyed or damaged during manufacture, inspection, assembly, sorting, or shipping. The failure of vital equipment in the field could endanger human lives or impair critical activities. In addition to the loss of service of the equipment until repairs can be made, the cost and time lost while diagnosing the malfunction and replacing the defective circuit board or module can be expensive. Later, of course, the faulty device must be replaced in that board—also a time-consuming, labor-intensive task.

The physics underlying ESD are as well known as the principles for protecting devices and circuits from its ravages. However, damage or destruction of electronic components or systems by ESD is, at heart, a *people problem*. Rising concern about the damage wrought by ESD has led to the inclusion of protective circuits at both device and circuit-board levels. These circuits will raise the lower thresholds of damage or destruction, but will not eliminate the problem.

On the positive side, more companies that handle or use ESD-sensitive devices have adopted ESD-control programs that include personnel training. More manufacturers are offering conductive work surfaces and flooring, personal grounding systems, packaging, containers, tools, and

Introduction

other accessories required for a well-planned, cost-effective ESD-control program. Semiconductor manufacturers, distributors, OEMs, and field-service personnel have adopted increasingly standardized methods of ESD protection. Therefore, the prices of the special products and services have become more competitive.

In many cases, ESD control has been merged with procedures to eliminate or reduce electromagnetic interference (EMI) and the threat of damage by power line transients.

The ESD threat is greatest whenever personnel are handling unprotected, sensitive devices and assemblies for shipment, storage, assembly, inspection, or inventory. People can easily acquire a static charge and pass it to the hapless device. However, the devices also can be damaged by the discharge of static built up on nonconductive surfaces or by the sliding action of the devices on insulating materials. Fortunately, this threat can be reduced through proper control of the environment and attention to handling the devices.

ESD transients have very fast rise times, very short durations, and amplitudes that can exceed thousands of volts. These characteristics make it difficult to provide complete protection for sensitive devices, including those with internal or external suppression networks that provide only marginal protection.

ESD control calls for eternal vigilance, training, and frequent refreshing of all personnel who come in contact with ESD-sensitive products. Unfortunately, ESD protection is a moving target because of the increasing vulnerability of devices, particularly very large scale CMOS and gallium arsenide (GaAs) ICs. Increasing gate densities, finer line structures, and thinner dielectric layers all perpetuate the ESD problem. Thus, the search for total ESD protection is seemingly never-ending; apparently it will not be eliminated by changes in semiconductor technology in the foreseeable future.

This book is a practical introduction to ESD control. It explains the origins and effects of ESD, and the methods, materials, and equipment needed to reduce these effects. These include instruments for measuring, monitoring, and controlling static, and equipment for conditioning the environment and simulating ESD under controlled conditions. Topics include: protective packaging, air humidification and maintenance of ion balance, and selection of appropriate tools and work surfaces.

This book gives you a basic background knowledge of electrostatics; the causes and effects of ESD on components, assemblies, and systems; and the concepts behind protecting sensitive devices from direct contact with the static discharges, as well as fields created by those discharges.

This book should be of value to everyone involved with the manufacture, assembly, quality control, testing, or repair of ESD-sensitive

Introduction

components or equipment. It should be especially useful to anyone who supervises or manages those activities. This book provides current knowledge of ESD and control for those who "need to know," although they might never handle sensitive products.

Chapter 1 defines electrostatic discharge and explains why anyone who works on or near semiconductor devices and circuits is a potential threat to those products. It relates ESD to other phenomena caused by static electricity.

Chapter 2 gives the principles of electrostatics so you can understand the physics of ESD and the principles behind its control in the electronics industry.

Chapter 3 discusses the damage mechanisms in ESD-sensitive devices, emphasizing the latest high-density ICs. It discusses how ESD can cause instant destruction or latent damage. It also considers the effects of these mechanisms on circuit boards and host equipment.

Chapter 4 covers the ESD-control networks being built into ICs, as well as suppressors and filters put on circuit boards and mounted within equipment enclosures. It also discusses the effects of shielding in equipment enclosures.

Chapter 5 explains special materials and describes conductive and static-dissipative containers, including trays, boxes, and bags used for protection during manufacture, inspection, assembly, shipment, and storage.

Chapter 6 discusses the humidity controls, work surfaces, flooring, air ionizers, tools, and production equipment recommended for use in ESD control. It also discusses the application and function of antistats.

Chapter 7 explains grounded wrist and foot straps, as well as optional protective clothing used in personal ESD protection.

Chapter 8 describes test equipment specifically designed and built to detect the presence of static charges and measure their magnitude and polarity. It also discusses the instruments for measuring the properties of materials related to static generation and dissipation.

Chapter 9 discusses various ESD-simulation equipment intended to meet differing test objectives. The design and construction of ESD simulators are dictated by the standards that are mandated or that reasonably apply to the testing and characterizations of products.

Chapter 10 presents a complete program for ESD control to be used as a guide in the preparation of plant specifications. Included are sections on the static-free workstation and handling procedures for ESD-sensitive products.

Appendix A is a listing of known commercial vendors offering ESD control products including antistats, clothing, grounding systems, ionizers, materials, packaging, test instruments, flooring, and workplace

Introduction

equipment. Addresses and telephone numbers are provided along with codes designating their areas of participation.

Appendix B contains the names and addresses of important agencies in the ESD field and a list of references used in the preparation of this book.

1
Static Electricity: A Growing Menace

Most electrostatic discharge (ESD) destruction and damage to semiconductor devices is caused inadvertently by one inspecting, sorting, or installing the devices. Destructive levels of ESD can be so weak that they are imperceptible even to people who know about the threat of ESD; they might expect to feel a slight sting or see a spark, as happens after walking across a carpet on a cold, dry day and touching a metal object.

Thus, ESD is an unseen and unfelt enemy that can strike wherever static charges are permitted to accumulate and whenever the person handling the sensitive device is not properly grounded. Perpetual vigilance is the key to effective ESD control, as is discussed later in this book, a momentary lapse of caution in even the best equipped workstation can defeat the most elaborate protection.

Although ESD damage can occur as a result of static buildup on a machine or even on a sheet of paper or plastic cup, human body static discharge is its most likely cause. Walking, scuffing one's shoes on the floor, or shifting around in a chair can build up a charge on the human body which acts like a capacitor storing energy. The discharge of this energy through the finger, hand, or through a metal tool held in the hand can zap a device and *destroy* it.

With such precautions as the use of conductive work surfaces, containers, grounded wrist straps and other precautions, one can prevent himself from becoming an agent of destruction. Maintenance of the relative humidity in the work area within specified limits, and neutralization of ion imbalance on nonconductive surfaces are other important steps in an ESD control program. This book considers all of these preventive measures in detail.

Static Electricity: A Growing Menace

Even as more companies and people are learning how to control static buildup and prevent ESD, the devices are becoming more vulnerable, making ESD protection a never-ending catch-up activity. Semiconductor manufacturers are under continuous pressure to design and add more functions on a chip, resulting in more transistors and elementary circuits called *gates* per unit area of silicon substrate. In addition, all designers are working toward higher speed and lower power consumption per device, particularly the very-large-scale integrated (VLSI) circuits. This trend makes CMOS and gallium arsenide devices even more susceptible to destruction than the earlier generation bipolar devices.

Even if the devices are not destroyed during handling, they could be damaged or impaired. Certain kinds of latent damage cannot be detected with routine tests and it could show long after the device is installed in its host equipment. This can occur even if the host equipment has not been electrically stressed; therefore, destroyed or damaged devices must be replaced. The sooner the damage or destruction is detected, the lower the cost of replacement. If the customer or user can demonstrate that the device was "dead on arrival" and not damaged or destroyed during unpacking, it could be returned to the vendor for replacement. Nevertheless, there might still be a loss of labor and time spent inspecting the device. Even if there is no direct cost of replacement, there can be intangible costs related to back ordering and production delays that are not easy to identify and plot. High losses of a part as a result of ESD might require an investment in surplus inventory. Increased inventory and the use of floor space to store parts are additional cost factors.

If the damaged or destroyed device is first detected at the circuit-board level, the unit price of the part is added to the cost of assembly labor lost. It has been estimated that the direct cost of replacement at this level is between $15 and $20 today.

If the damage or destruction of the device occurs after the product or system has been assembled, the direct cost of replacement rises to $150 or more. The defective part must be unsoldered and replaced and the circuit board retested. At this level, the indirect cost could include penalties in late shipment and disrupted production schedule.

It has been estimated that the cost of replacing a failed or damaged component in the field could be as high as $1,500. The services of the product or system are lost during the time it takes to replace the faulty host circuit board or module. Thus, the cost of replacement rises exponentially as the device moves through the progressive stages of inspection and assembly into the finished product. This situation is illustrated graphically with the square law curve in FIG. 1-1.

In end products such as medical or life-support equipment, failure of the system could result in loss of life; in process control equipment, it could cause property damage. If the host equipment is a production

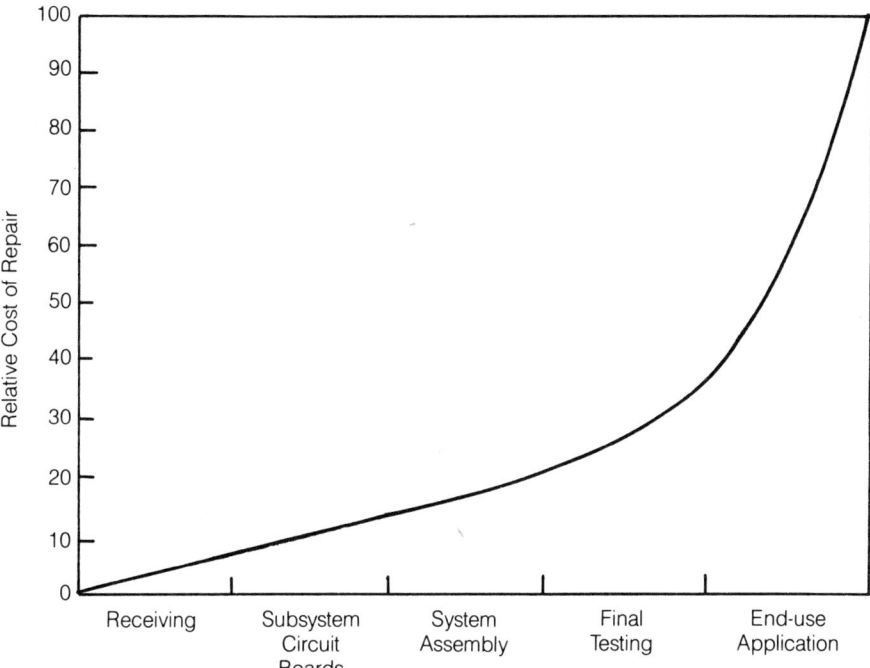

Fig. 1-1. Estimated cost of component failure as a result of ESD at progressive steps in end product manufacture.

machine, its downtime could cripple production; if it is data processing equipment, it could disrupt normal office operations and even result in lost data.

A field failure can also call into question the manufacturer's quality and test procedures, to say nothing of its reputation in the industry. Even after the faulty device has been traced and the host circuit board replaced, the circuit board must be returned to a service shop or the factory for component replacement at the board level.

The higher the function or gate density of the integrated circuit (IC) and the finer its line structure, the more vulnerable it is to the ravages of ESD, now recognized as the most common cause of electrical overstress. Moreover, as this book points out, architecture of the device and the thickness of its dielectric layers have a significant bearing on its ESD susceptibility.

Gate density in ICs is moving toward a goal of 1,000 gates per square millimeter; propagation delays are decreasing toward the nanosecond level; and speed-power products are decreasing toward 1 picojoule. Moreover, supply voltages are decreasing from 5 volts to 2 and 3 volts; channel lengths are decreasing toward 1 micron; and line dimensions are closing in on 0.05 micron.

Static Electricity: A Growing Menace

Lack of chip-level protection can leave ICs vulnerable to damage by transients as low as 50 volts—a value well below the threshold of human perception. Depending on circumstances and the individual, the threshold of ESD perception is considered to be between 3,000 and 4,000 volts. Semiconductor manufacturers, aware of this problem for many years, have been building protective networks into their products.

Some IC makers claim they have raised the level of ESD susceptibility of their devices from under 100 volts to 2,000 volts. Even built-in protective networks do not provide complete protection because they can be damaged or destroyed by voltages over 2,000 volts, and still be below the human perception level. The networks might provide only "one-shot" protection; and a component could be destroyed by the first overvoltage encountered. After that, the device is no longer protected, even from a lower second transient. Most devices are routinely exposed to ESD voltages well in excess of 2,000 volts throughout the entire assembly and test process, so they must be adequately protected.

Many original equipment manufacturers (OEMs) have supplemented on-chip protection with protective networks and filters mounted on the circuit board. These include metal-oxide varistors (MOVs) and silicon transient voltage suppressors (TVSs). TVSs are specialized forms of zener diodes. But even these external networks do not ensure complete safety of the device; while destruction can be prevented, latent damage still could occur. Thus, even if the device reaches the field without detectable damage, it can be the cause of an annoying operational upset in the host equipment. If the host equipment is a digital computer, it might have to be turned off then reset to clear a temporary fault caused by the defective component.

ESD was rarely the cause of electronics failures during the vacuum tube era. The high-voltage handling capability of vacuum tubes and the higher voltage ratings of components of that era effectively precluded ESD problems. Even in the next evolutionary step to discrete transistor circuits, ESD still was not a serious threat. However, integrated circuits became the first casualties of undetected ESD before they left the factory. Damaged devices were detected as a result of mysterious malfunctions of computers, television sets, and other electronic equipment redesigned around ICs. Before long it was found that passive components designed to operate at the lower voltage levels of ICs also were becoming ESD casualties.

These mounting losses led to changes in the workplace—the grounding of persons who handle ESD-sensitive devices, the grounding of tools, and the use of conductive containers for shipping and temporary in-process storage. But, the most important lesson learned was the need for modification in personnel behavior. The use of wrist or shoe ground straps was mandated for all persons handling ESD-sensitive devices or assemblies containing them. Training programs were introduced to raise the consciousness of all who handled electronic devices and circuit boards.

Unfortunately, the rapid spread of ESD-sensitive electronics into virtually all walks of business and industry outdistanced the spread of knowledge of the importance of ESD control. To this day, needless damage and destruction of devices continues to occur, literally at the hands of untrained people in semiconductor manufacturing plants, at distribution facilities, and in the OEM's factories.

Concern for the protection of equipment against the destructive effects of ESD was soon reflected in the redesign of equipment enclosures, the selection of more appropriate materials, and the use of shielded cables and connectors. In many cases, provisions for protection against ESD have been merged with those for protection against other overvoltages, including electromagnetic interference (EMI) and nuclear electromagnetic pulses (NEMPs).

WHAT IS ELECTROSTATIC DISCHARGE?

When two dissimilar materials are brought in contact by rubbing or are separated rapidly, one tends to attract electrons away from the other. As a result, each develops a different charge voltage level. Certain combinations of materials when rubbed together or separated can achieve differentials of thousands of volts.

Most people have had experience with ESD on cold, dry winter days when the relative humidity in the room is less than 40 percent. Simply walking on a carpeted floor and touching a metal doorknob or key can result in a stinging shock on the finger closest to the metal object. Under reduced lighting conditions it is possible to see the spark between the finger and the metal object. Although startling and annoying, the experience is usually not harmful to most people.

As one walks, the materials in the carpet and in the soles of the shoes become oppositely charged. The charge on the shoe soles induces an opposite charge in the human body, which acts like a capacitor. When an uncharged doorknob or key is touched, the human "capacitor" discharges.

A similar experience can occur when a person slides across the seat of an automobile and then touches a metallic part of the car during cold, dry conditions. Nevertheless, in both examples, it is important to note that if the shock can be felt and an arc seen, the voltage is well above that needed to "kill" an ESD-sensitive semiconductor device.

Measurements of ESD created by people in these situations indicate that peak discharge currents can reach tens of amperes. A maximum peak voltage level of 15,000 to 20,000 volts can be attained. A person charged to 20 kV can produce an ESD spark up to 3/4 inch long.

Far less dramatic but nevertheless lethal ESD for sensitive electronics can strike from wooden, plastic, paper, or other nonconducting surfaces. The static electricity buildup on these objects is far lower than that necessary to form an arc that can be perceived by the human eye. The "clinging" of clothing removed from a drying machine is a common demonstration of static buildup on nonconducting materials.

ESD can also be transmitted by metal furniture, such as chairs, shelves, or instrument carts. However, the sharp corners and edges of metal furniture tend to bleed off a charge in the form of a corona, holding the maximum voltage expected from furniture to only 6,000 to 8,000 volts. Since metal furniture is more conductive than people or nonconductors, higher peak discharge currents occur at the same charge voltage. Peak current amplitudes of 170 amperes have been measured.

The human body can be charged to a higher voltage than a metal chair, so its stored energy can be higher, as much as 20 to 30 millijoules (mJ) versus 5 to 6 mJ for metal furniture.

The upper frequency of ESD can reach the 1- to 5-GHz microwave region. At these frequencies equipment cables and conductors on circuit boards become efficient receiving antennas. As a result, ESD can induce high levels of voltage in both analog and digital equipment.

The electrical energy associated with ESD can enter electronic equipment either by conduction or radiation. When the source is within approximately 12 inches of the sensitive device, the primary form of radiated coupling can be either capacitive or inductive, depending on the impedances of the source and receiver. At greater distances electromagnetic field coupling predominates.

Destructive damage to semiconductor devices, discrete transistors, diodes, and ICs has two primary causes: heat created by ESD current at constriction in devices such as bipolar transistors, leading to secondary breakdown; and the high voltage differentials leading to dielectric breakdown in MOS transistors. Both damage mechanisms can occur in a single device. For example, dielectric breakdown can trigger high current flow, which causes thermal failure. A third form of failure is overcurrent in conductors, leading to burnout and an opening of the metallic paths. These damage modes are discussed in greater detail in chapter 3.

The voltages and currents that cause permanent damage in semiconductor devices are one to two orders of magnitude greater than those that cause an equipment upset. Damage is more likely the result of the ESD arc directly in contact with device pins or circuit board lines; equipment upset is normally caused by radiated coupling.

ESD-induced voltage and/or currents can cause an upset in circuit operation if they exceed the signal levels in the electronic circuit. In high-impedance circuits, little current is drawn and the signals are at voltage levels; capacitive coupling dominates and ESD-induced voltages are the most likely culprit. By contrast, in low-impedance circuits the signals are current, so inductive coupling dominates. Most problems are traceable to ESD-induced currents.

ELECTROSTATIC DISCHARGE AND ELECTRONIC DEVICES

Advanced VLSI devices now have more than 1 million transistors on a chip, but the IC manufacturers continue to crowd on more and more. At the same time, they are increasing the operating speed of ICs and reducing their power consumption.

The most widely used technology for the fabrication of ICs today is metal-oxide semiconductor (MOS) technology, and the basic building block is the MOS field-effect transistor (MOSFET). To reach higher component density, higher speed, and lower power in MOS technology, manufacturers are making the gate oxides (dielectrics) of the thousands of on-chip transistors thinner, junctions shallower, and interconnecting lines finer.

Although electrostatic discharge has been a problem with MOSFETs from the time of their invention, the problem is becoming worse with each increase in the complexity of a chip. The thinner the oxide, the lower the breakdown voltage. Because of the severity of the problem, reliability engineers describe ESD as our most significant electronic component problem.

For example, PMOS and NMOS ICs have channel lengths of approximately 10 microns (1 micron = 0.000001 meter) and an oxide thickness of 1,100 to 1,500 angstroms (1 angstrom = 0.0001 micron). Devices now being developed have channel lengths of $1/4$ micron and dielectrics that are only 150 angstroms thick.

However, ESD is also a problem for many devices other than MOSFETS and MOS ICs. It has taken its toll on bipolar transistors and some emitter-coupled logic (ECL) and transistor-transistor logic (TTL) ICs. It has also caused the destruction of surface acoustic wave (SAW) devices, operational amplifiers, JFETS, SCRs, microwave transistors and diodes, thin-film resistors, Schottky diodes, resistor chips, and piezo-electric crystals. In some cases, even relays, connectors, and printed circuit boards have been victims.

Once installed in a circuit board or end-use equipment, an ESD-sensitive device can be less vulnerable that it was as an individual component. However, it could be more vulnerable if the conductive paths or traces on the circuit board act as antennas to gather even greater voltages to destroy the devices. As a result, it is best to assume that all circuit boards and subassemblies containing ESD-sensitive components are vulnerable to ESD damage.

There are no precise figures on the worldwide cost of ESD destruction to electronic components and products. Estimates range as high as $1 billion per year. This seems believable when you consider the geometric growth in the production of all kinds of semiconductor devices in the world each year and the cost of finding and replacing installed devices destroyed or damaged by ESD.

It has been estimated that between 5 and 25 percent of all component failures are caused by ESD and it is seen as the cause of 50 percent of all "dead-on-arrival" components. Statistics indicate that more than 50 percent of early operating failures are caused by ESD. One manufacturer estimated that 60 percent of his field service calls were related to ESD damage.

ESD is a problem for everyone in the electronics industry, from the device maker and the original equipment manufacturer to the service technician and the user who operates the equipment in the field. ESD also affects consumer electronics from high-fidelity equipment to television sets and video cassette recorders.

ELECTROSTATIC DISCHARGE AS A UNIVERSAL PROBLEM

Although some facilities and locations have more severe problems with ESD than others, it is a universal problem for everyone engaged in any form of electronic equipment manufacture, test, or repair. It is tempting to deny the existence or magnitude of the ESD problem because:

- Static discharge that can damage or destroy ESD-sensitive devices and circuits cannot be seen or felt.
- Many believe that ESD is only a problem with MOS transistors or CMOS ICs.
- Many believe that the ESD-protective networks built into the ICs provide adequate protection.
- Many believe that once a sensitive component is installed on a board or in a module there is no longer an ESD problem.
- Failures as a result of ESD can be attributed to other causes.
- By denying an actual or potential problem with ESD, the cost of ESD protection, control training, and time lost carrying out the procedures can be saved.

The instruments described in chapter 8 include those intended to determine the magnitude of electrostatic problem voltages present. Actually, no tests are necessary to show the existence of harmful static electricity in a work area. These voltages are detectable in nearly all work areas around the country at some time during the year.

Even when potentially dangerous static charges are measured in a room, some people could blame destruction of known ESD-sensitive devices on power line transients or improper assembly or testing. Unfortunately, the only sure way to prove that ESD damage has occurred is to send a failed device to a specially equipped laboratory for failure analysis. In this procedure the chip is removed from its package and examined under a microscope for clues. Because conventional binocular microscopes lack sufficient optical resolution, the scanning electron microscope (SEM) is normally employed.

The SEM uses electrons, rather than light beams, to illuminate objects, and it enables one to view the sites of destruction with high levels of magnification. For example, it permits the viewing of the tiny pinholes that ESD has punched through the dielectric or the melted and fused conductive paths caused by ESD. However, these microscopes are expensive instruments that are generally affordable and cost effective only for testing laboratories and large semiconductor corporations.

Even with the detailed view of damage sites made possible with the SEM, it is difficult to verify with 100 percent accuracy that ESD was the cause of visible damage. As an alternative to SEM examination, practical tests called *A/B comparisons* can be run in the factory. Lot A is handled by normal procedures; lot B, run at the same time, is handled with all the ESD precautions. If the lots are large enough and the tests sufficiently discriminating, the difference in yields is apparent. Also, a comparison of winter and summer yields generally reveals the cyclic pattern associated with ESD—higher failure rates in the winter when the relative humidity in the plant is lower.

Although some plants might not experience device or product failures attributable to ESD at the factory, still there could be undetected problems in the field. If field engineers and technicians are not trained to recognize failures caused by ESD, they will not report them back to the factory. The only way to avoid this situation is by insisting on periodic in-depth analysis of field failures.

A complete and satisfactory solution to ESD problems can be both costly and time-consuming, and companies that have escaped the problem so far could be faced with it in the future as IC density increases. The recommended ESD control program can call for a reorganization of all materials-handling practices within the plant. It could also call for sweeping changes in employee work habits and the enforcement of rules for the wearing of grounded wrist straps and appropriate clothing. It might also necessitate structural changes in the interior of the building, including the resurfacing of floors or walls or the installation of adequate heating, ventilating, and air humidification.

Static Electricity: A Growing Menace

An ESD-controlled area or room should be designated for the handling of sensitive semiconductors and circuit boards. Also, the installation of conductive benchtops could be necessary. Conductive tote boxes, trays, and packaging materials should be required, if they are not already in use, and restrictions should be placed on traffic through the ESD-controlled area.

Static awareness, a subject discussed in detail later in this book, must be encouraged, and new and seasoned employees alike reminded of the consequences of lapses in established routines. Some plant or department managers could reach the conclusion that the cure is worse than the disease. But most evidence suggests that there is a short payback on the cost of an ESD control program. Chapter 10 is a guide for the establishment of a plantwide ESD program that can be tailored to fit specific conditions.

A SHORT HISTORY OF STATIC ELECTRICITY

Electrostatic discharge in one form or another is the earliest known form of electricity. The most dramatic and dangerous form of ESD is lightning, and it has awed mankind throughout recorded history. Lightning as a destructive force has killed or injured human and animal life as long as they have existed on earth. Ironically, many scientists believe that life on this planet came about through a synthesis process powered by the discharge of static electricity in a primordial mix of chemicals.

Aside from damage by direct strike, lightning has been responsible for secondary fires that have also taken their toll of life, forests, and crops. As man became more civilized and built homes, factories, and other flammable and unprotected structures, losses from lightning strikes mounted. Finally, countermeasures in the form of fireproof structures and lightning rods were developed to reduce these losses.

However, other forms of man-made ESD resulting from friction and separation of materials have been blamed for explosions in chemical plants, hospital operating rooms, refineries, mines, grain elevators, and other places where explosive gases are released or accumulate.

Not all static problems result in explosions or fires. In many manufacturing processes, static charges interfere with production. Powders resist cohesive mixture, textile fibers twist or break when being drawn, and paper and plastic film cling and resist movement. Some of the causes include the use of high-speed production machines, the proliferation of synthetic materials, and wider use of very thin paper and film.

Because static electricity attracts dust, it can cause defects in surface finishing or polishing of products, affecting their quality. The causes and cures of these industrial ESD problems are, however, beyond the scope of this book.

A Short History of Static Electricity

Nevertheless, static charges have positive benefits. They have been employed effectively in copying machines based on Xerography and in laser printing. Static charges are employed in the graphic arts industry to make direct and indirect screens without the need for a vacuum frame, in metal working as a holding force in chucks, and in the electrical industry to test the integrity of cable insulation.

By 1960, researchers had discovered that ESD could damage semiconductor devices, particularly integrated circuits. In 1966, the Richmond Corporation developed antistatic pink polyethylene as a result of a disastrous premature missile ignition at Cape Canaveral, which was attributed to ESD. This material was later to be used widely for packaging electronics and for use in ESD control.

By 1975, the Jet Propulsion Laboratory and Hughes Aircraft Corporation, to name two organizations, had prepared draft procedures for the protection of microcircuits from ESD. Within two years, seminars on ESD control were being conducted by the Reliability Analysis Center. In 1979, the first annual Electrical Overstress/Electrostatic Discharge Symposium sponsored by the ITT Research Institute was held.

In 1980, the Department of Defense (DOD) published two important and significant documents: DOD Handbook 263 (DOD-HDBK-263) and DOD Standard 1686 (DOD-STD-1686). These references were prepared by the U.S. Naval Sea Systems Command to provide guidance in developing, implementing, and monitoring elements of an ESD control program.

The EOS/ESD Association was formed in 1982 with its objective to "Strive for advancement of theory and practice of electrical overstress avoidance . . . The field of interest of the Association shall be the design hardening and prevention aspects of electrical overstress. This especially includes phenomena of electostatic discharge and its control as applicable in design, manufacturing, and end use." The membership of the EOS/ESD Association now represents most of the major electronics manufacturers in the United States.

THE ESD ENVIRONMENT

ESD damage is much more likely to occur in dry than wet climates. For example, the ESD problem is likely to be more severe in Phoenix, Arizona, with its typically low outside relative humidity of 12 percent than in cities like Los Angeles, where the outside relative humidity rarely drops below 45 percent.

When heating systems are turned on in the fall and winter, inside room humidity can drop to levels below even those of outside readings in Los Angeles, where relative humidity readings of 25 percent are not uncommon. Under these conditions, component losses traceable to ESD increase.

Static Electricity: A Growing Menace

Fig. 1-2. Electrical transients are produced by many sources.

The ESD Environment

Semiconductor devices, sensitive passive components, and circuit boards must survive in a hostile environment where there can be many different sources of transient electrical overstress. As shown in FIG. 1-2, these transients include nuclear electromagnetic pulses (NEMPs), lightning, fields from power transmission lines, interference caused by radars and radio communications equipment (EMI and RFI), as well as primarily man-made electrostatic discharge.

Figure 1-3 illustrates the bands of frequencies covered by these electrical transients. Nuclear electromagnetic pulses present the widest bandwiths, blanketing a large part of the frequency spectrum with high field-intensity interference. The interference from lightning, although also at high field intensity, is largely confined to the lower frequency regions. Communications and radar systems emit lower field intensities at different parts of the frequency spectrum.

The duration of some of these transients can be measured in time values as short as a few nanoseconds or as long as milliseconds, as shown in FIG. 1-4. The fastest of the transient waveforms is produced by NEMPs, and the slowest by lightning strokes. The other waveforms fall between these two limits if plotted on the same time scale. ESD is at least as fast as NEMP and has fundamentally the same wave form, although it does not have as high an amplitude. One of the important findings of ESD research is that the characteristics of device failures depend on the amplitude and duration of the overstress transients.

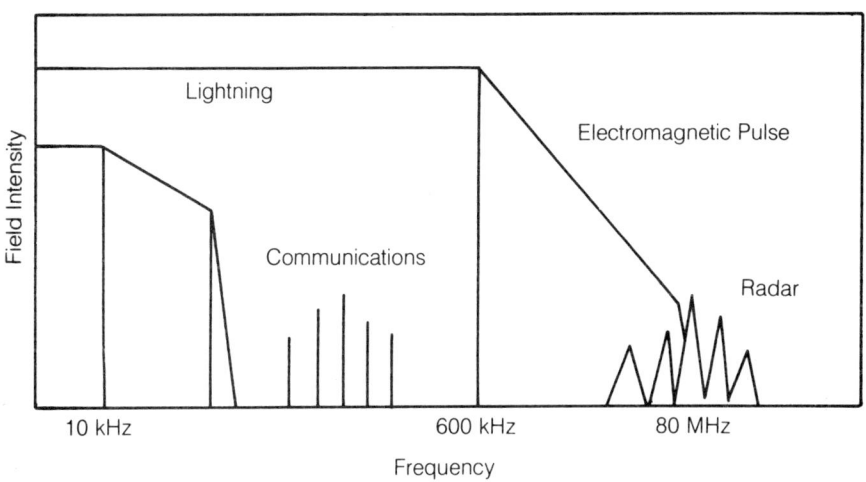

Fig. 1-3. The spectra of various sources of electrical transients.

Static Electricity: A Growing Menace

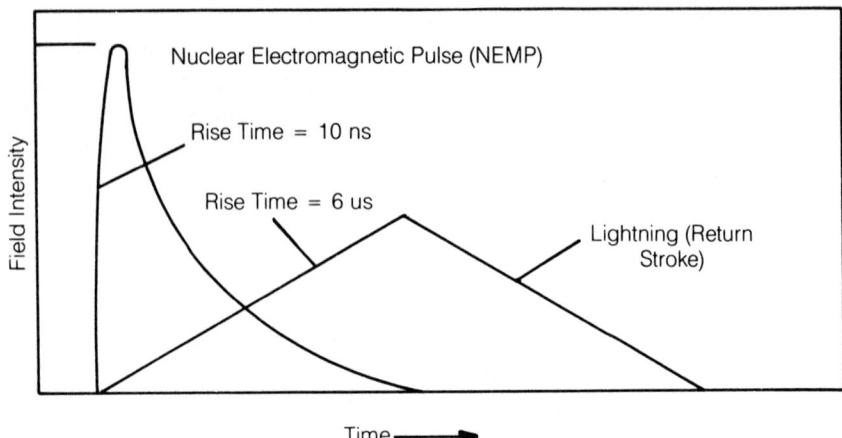

Fig. 1-4. Waveforms of various sources of electrical transients. Nuclear electromagnetic pulses (NEMP) are fastest and lightning strokes are slowest.

2
Principles of Electrostatics

Static electricity is defined as an excess or deficiency of electrons on a surface. It is produced when two materials meet, either through rubbing or simple contact, and then are separated. A typical example occurs when two electronic components slide around in plastic shipping containers. After the components are separated, one will have a positive electrostatic charge, the other a negative charge. The polarity and size of the charge are determined by the speed and duration of the sliding motion and the material composition of the devices.

Static electricity is also generated by a person walking across a carpet, sliding off of upholstered chairs, or rubbing his or her arms across the tops of a workbench. Static charge levels as high as 25,000 volts (V) have been measured on persons under conditions of very low relative humidity. Even at 55 percent relative humidity, charges as high as 7,000 volts have been generated by human body movement.

An understanding of the tools and techniques for controlling electrostatic discharge calls for a knowledge of the basic principles of *electrostatics*, the science of electric charge at rest. Long before electrostatic discharge became a threat to electronics devices and products, static electricity was a subject for study in elementary physics courses at the high school and college level. The study of static electricity usually precedes the study of current electricity and magnetism.

A knowledge of static electricity helps to explain the structure of the atom and the more useful, and usually more interesting, current electricity. As mentioned in chapter 1, the discharge of static electricity is recognized as the cause of explosions in mines, munitions factories, flour mills, and hospital operating rooms. For many years static control programs had

Principles of Electrostatics

been carried out in paper mills and plastic film production plants. It first became a topic vital in the protection of electronic devices and circuits with the development of integrated circuits, particularly the MOSFET and more sensitive passive components. Now, knowledge of electrostatics and methods of controlling it is part of the training of all electronics engineers, particularly those designing MOS devices.

Serious attention was first given to the study of electrostatics more than 2,600 years ago, when some Greeks observed that pieces of *amber*, a hard yellow-brown translucent fossil resin, when rubbed with cat's fur would "pick up" bits and pieces of almost any light material, such as hair and scraps of paper. This phenomenon was largely ignored for the next 2,000 years until it caught the attention of the English physicist William Gilbert who, in 1600, studied it in great detail.

Gilbert showed that the same attraction phenomenon could be obtained by rubbing substances other than amber. To describe this finding he coined the word *electrified*, from the Greek word for amber, *elektron*. But it was not until the eighteenth century that much more was learned about *electrified bodies*, bodies with electric charges.

Electrostatics is generally concerned with *stationary electric charges*, those that are at rest or that occasionally move around at random. When these charges move, they form electric currents similar to other electric currents except that they are short-lived, seemingly instantaneous transients. Because of their short life and discontinuous nature, these currents cannot be put to practical use in operating electric and electronic products.

ELECTRON THEORY AND ATOMIC STRUCTURE

To understand the nature of electric charges, it is necessary to examine the atomic structure of matter. All matter is composed of very small particles called *atoms* (FIG. 2-1), the smallest division of an element that retains its identity as an element. Although extremely small, the atom can be subdivided into three subatomic building blocks: protons, electrons, and neutrons. It is convenient to think of these particles as arranged in the form of a solar system: electrons correspond to the planets and orbit the nucleus made up of protons and neutrons, which correspond to the sun. The nucleus contains most of the mass of the atom.

The electron is a very small fundamental particle with a mass of 9.11×10^{-31} kg and a negative electric charge of 1.60×10^{-19} coulomb. It has a mass of about $1/1840$ of the proton's, which is 1.673×10^{-27} kg. However, it has a negative charge equal in value but opposite in polarity, or sign, to the positive charge of the proton. The charge e of an electron is the smallest charge unit known to exist. All atomic charges are some integral multiple of this electron charge: $2e$, $3e$, etc.

Electron Theory and Atomic Structure

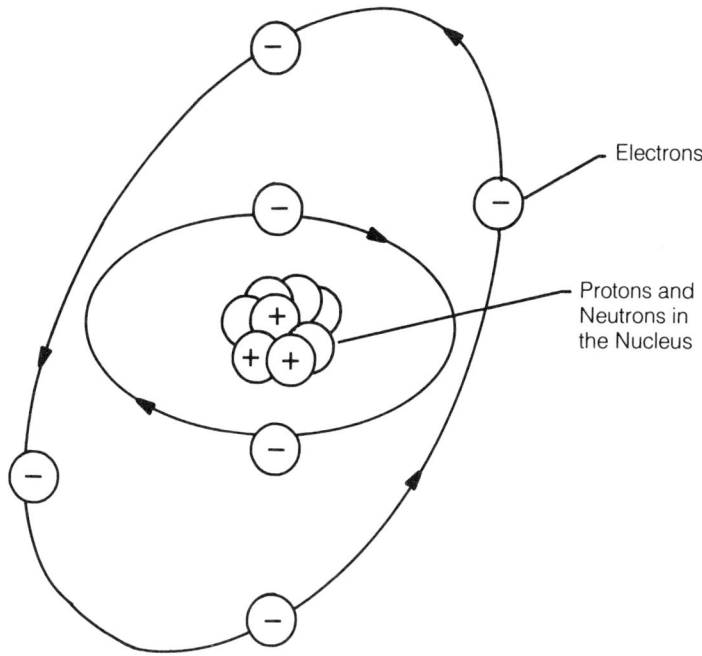

Fig. 2-1. Structure of the atom. Electrons orbit protons and neutrons in the nucleus.

Because the neutron has no charge, it is electrically neutral. But it has a mass slightly greater than that of a proton: 1.675×10^{-27}. Normally an atom will have as many electrons orbiting the nucleus as it has protons, and it is also electrically neutral. The orbits of the electrons describe imaginary concentric shells or layers around the nucleus; one or more electrons revolves in each shell. The electrons in the outermost shell— that is, the shell farthest from the nucleus—are capable of breaking away from the atom under certain conditions.

When electrons are removed from the atom, the neutrality of the atom is canceled and the atom takes on the charge of the positively charged nuclei; this give the atom a net positive charge. The atom that receives these electrons will then have an excess of electrons and a net negative charge. With this atomic structure in mind, it is easier to understand what happens when objects are electrified by rubbing or friction.

ELECTROSTATIC ATTRACTION AND REPULSION

When two electrified objects are brought close together, they either attract or repel each other. This phenomenon can be readily demonstrated with a simple experiment shown in FIG. 2-2. In this experiment a

Principles of Electrostatics

very light, small ball made of pith is suspended by a thread. A hard rubber rod is stroked with fur or flannel and then brought in contact with the ball. After a brief instant, the ball will jump away from the rod as if it were being repelled. Moving the rod near the ball again causes it to move away again. However, if a glass rod is rubbed with silk and brought near the ball, the ball will be attracted to the rod. Once it touches the rod, it is again repelled by the rod.

In the process of rubbing, the rod acquired an electric charge. When it touched the pith ball it transferred some of its charge to the ball. The ball then also became electrified. Because of this and other similar experiments, scientists concluded that there are only two different kinds of electric charges: the charge on the rubber rod rubbed with flannel is called *negative*, and the one on the glass rod is called *positive*.

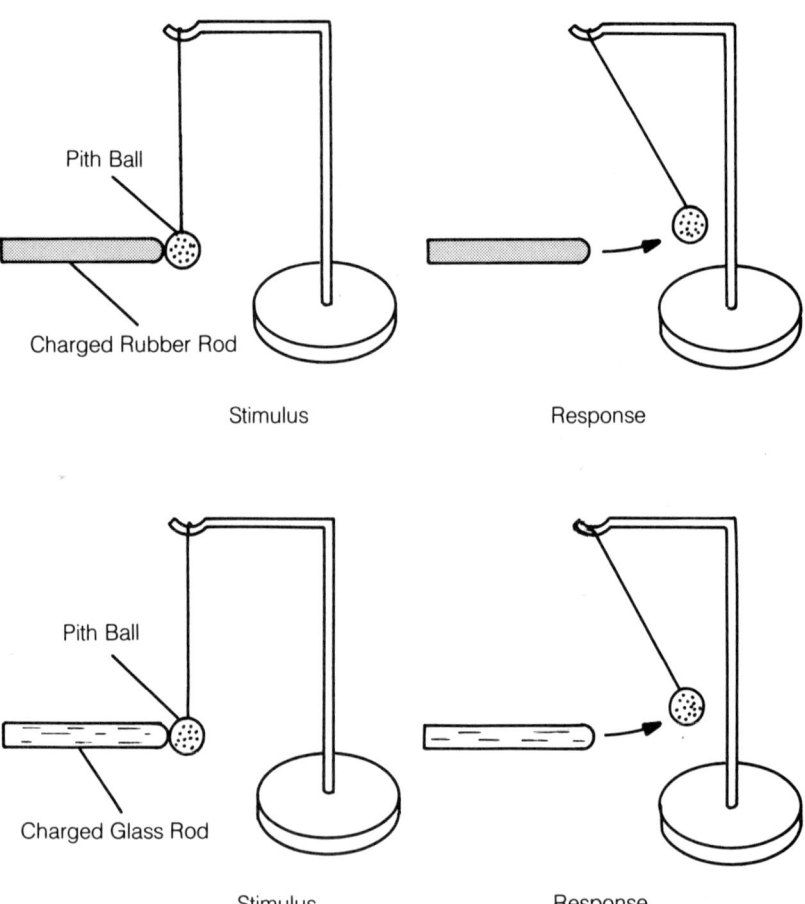

Fig. 2-2. Demonstration that like charges repel.

Electrostatic Attraction and Repulsion

When the positively charged glass rod touched the ball, it gave it a positive charge and the ball was repelled by the rod. But when the negatively charged rubber rod touched the ball, it gave it a negative charge and the ball was repelled by the rod. It can therefore be concluded that like charges repel one another. If a positively charged pith ball is brought near a negatively charged pith ball, the two balls will attract each other. Thus, unlike charges attract one another.

It is convenient to say that static electricity is created or generated. Actually, however, electric charge cannot be created; it is merely transferred. When the rubber rod in the experiment received a negative charge, the flannel received an equal and opposite positive charge. A neutral body has an equal amount of positive and negative charge; its net charge is zero.

Because unlike charges attract and like charges repel, it is easy to determine the charge on unknown bodies. If a body is brought near a charged glass rod suspended by a thread, the tendency of the rod to repel or attract the body can be seen. If it repels the body, the unknown body has a positive charge. However, if it attracts the body, the unknown body has a negative charge.

The force between charges can be found by the use of Coulomb's Law. Charles Coulomb (1736–1806) was a French physicist who introduced the concept of *point charges*, charged bodies who dimensions are small in comparison with the distance between them. He showed that the force, F, of attraction or repulsion between two point charges q_1 and q_2 is proportional to the product of the charges and inversely proportional to the square of the distance, r, between them. This is stated in equation form as:

$$F = \frac{q_1 q_2}{kr^2}$$

where: F = force in newtons
k = a dielectric constant depending on the medium between the charges.

All mediums other than a vacuum act to lessen the force; however, the difference between air and a vacuum in this equation is negligible. If the charges have like signs, the force repels the charges; if the charges have opposite signs, the force attracts the charges. Coulomb's Law is the fundamental equation of electrostatics.

The term *coulomb* has been given to the unit of charge equivalent to the charge carried by 6.24×10^{18} electrons. Since this is an unmanageable unit in the study of electrostatics, smaller units such as the microcoulomb (μC), the nanocoulomb (nC), and the picocoulomb (pC) are used. Note that 1 μC = 10^{-6}C, 1 nC = 10^{-9}C, and 1 pC = 10^{-12}C.

Principles of Electrostatics

TRIBOELECTRIC CHARGING

When movement occurs between two bodies, particularly if they are dissimilar, one of the bodies will tend to lose or give up electrons more readily than the other. In effect, electrons will be stripped or displaced from one body and transferred to the other. The body that has lost electrons will then have a positive charge, while the body that has received the electrons then has a negative charge.

The transfer of electrons takes place rapidly and then diminishes as the surface energies equilibrate. The generation of static electricity by this method is called the *triboelectric effect*. The word is derived from the Greek *tribein*, to rub, and *tribos*, the action of rubbing.

Each material has an electrostatic potential that can be measured with an electrostatic voltmeter. The voltage generated may be 100V to 35,000V, as shown in TABLE 2-1. Its magnitude depends on speed of movement or separation, types of materials, humidity, surface characteristics, and surface geometry.

The rubbing together of two materials is the movement that most often causes triboelectric charging. In the process of rubbing, the materials are repeatedly brought into close contact and then separated. Actually, rubbing is not necessary for triboelectric charging to occur. The act of separating two objects, even of the same material, that have been in contact can generate substantial electrostatic charges, as demonstrated by separating the sides of a plastic bag or unrolling transparent plastic tape (FIG. 2-3).

Rubbing or contact and separation can occur in many ordinary activities where no conscious effort is being made to generate static electricity. These actions include walking across a floor, removing a coat or sweater, standing up from a seated position in a chair, removing or placing an object in a container, removing an IC from a plastic envelope, and moving ICs in dual-in-line package (DIP) cartridges or tubes during shipment. TABLE 2-2 lists other typical prime sources of charge.

Table 2-1. Typical Electrostatic Voltages (DOD-HDBK-263).

Method for Generating Static	Values in Volts	
	10 to 20 % Relative Humidity	65 to 90 % Relative Humidity
Walking across carpet	35,000	1,500
Walking over vinyl floor	12,000	250
Person at bench	6,000	100
Open vinyl envelope for work instructions	7,000	600
Pick up polyethylene bag from bench	20,000	1,200
Sit in chair padded with polyurethane foam	18,000	1,500

Triboelectric Charging

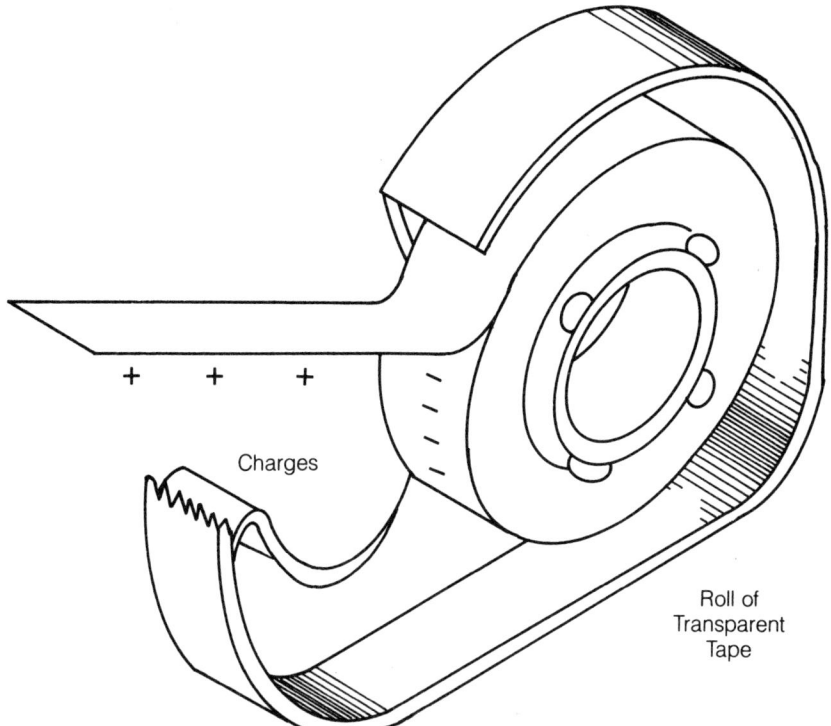

Fig. 2-3. Generation of static electricity by separating two objects.

The types of materials brought together will determine the magnitude and polarity of the charges. These factors are related to their position in the *triboelectric series*, a list of materials in an order of positive (+) charge to negative (−) charge as a result of the triboelectric effect. A material higher on the list, shown in TABLE 2-3, is positively charged when rubbed with a material lower on the list.

The position, or order of ranking, of materials in the triboelectric series should be considered only as approximate. Many different triboelectric series have been published and the positions of some materials differ. However, the value of the series is the knowledge that, the farther apart two materials are in any series, the greater is the magnitude of the charge created during triboelectric charging. Charge magnitude can be affected by surface cleanliness, humidity, lubricity, and the amount of surface area involved in the rubbing action.

TABLE 2-3 shows that cotton is at the center because it is relatively neutral. The materials above it in the table are increasingly positive; those beneath it, increasingly negative. A material can have either a positive or negative charge, depending on what it is rubbed with.

Principles of Electrostatics

Table 2-2. Typical Prime Charge Sources (DOD-HDBK-263).

Object or Process	Material or Activity
Work surfaces	Waxed, painted or varnished surfaces Common vinyl or plastics
Floors	Sealed concrete Waxed, finished wood Common vinyl tile or sheeting
Clothes	Common clean room smocks Common synthetic personnel garments Nonconductive shoes Virgin cotton*
Chairs	Finished wood Vinyl Fiberglass
Packaging and handling	Common plastic—bags, wraps, envelopes Common bubble pack, foam Common plastic trays, plastic tote boxes, vials, parts bins
Assembly, cleaning, test and repair areas	Spray cleaners Common plastic solder suckers Solder irons with ungrounded tips Solvent brushes (synthetic bristles) Cleaning or drying by fluid or evaporation Temperature chambers Cryogenic sprays Heat guns and blowers Sand blasting Electrostatic copiers

*Virgin cotton can be a static source at low relative humidities such as below 30 percent.

The larger the contact area between the two materials, the more electrons that can participate in the transfer. Thus, the magnitude of the triboelectric charge is proportional to the contact area. By minimizing the contact area through the use of ribs or standoffs, the charging of components and ICs in shipping tubes can be reduced.

Because triboelectric charging is a friction process, it can be reduced by increasing the materials's lubricity, a measure of surface smoothness and frictional characteristics. The higher the lubricity of the surfaces being rubbed, the lower the friction and, hence, the lower the generated charges. Lubricity can be increased through the use of materials called *antistats*, discussed in detail in chapter 6.

The charges developed during triboelectric charging are either mobile or immobile. The charges on conductors are *mobile*—they are rapidly dis-

Triboelectric Charging

Materials
Increasingly positive ↑
Quartz
Glass
Mica
Human hair
Nylon
Wool
Fur
Lead
Silk
Aluminum
Paper
Cotton
Steel
Wood
Amber
Sealing wax
Hard rubber
Nickel, copper
Brass, silver
Gold, platinum
Sulfur
Acetate rayon
Polyester
Celluloid
Orlon R   Saran R
Polyurethane
Polyethylene
Polypropylene
PVC (vinyl)
Silicon
Teflon R
↓ Increasingly negative

Table 2-3. Triboelectric Series (EIA-541).

tributed over the conductor's surfaces and the surfaces of other conductive objects they touch. The charges on insulators are *immobile*—they tend to remain in the localized area of contact. Since immobile charges are not readily distributed over the entire surface of the material, they generate high electrostatic voltage levels on insulators.

Electrostatic charges can be transmitted readily from an object to the conductive sweat layer of a person's skin, causing that person to be charged. When a charged person handles or comes in close proximity to an ESD-sensitive component or device, the device can be damaged by direct discharge (by touching the part) or by the creation of an electric field around that part.

Principles of Electrostatics

It is a basic law of nature that the net electric charge of any isolated (insulated) system remains constant. Charges can only be separated or combined; they cannot be created. Thus, in triboelectric charging, when charges are transferred from one body to another, for every negative charge there is an equal positive charge. In other words, what one body gains, another loses.

CONDUCTORS AND INSULATORS

In some of the earliest experiments with electrostatics, it was found that charges could be transferred easily from one point to another by metal conductive wires. Thus, metals, regardless of shape or configuration, were found to be good conductors of electrical charges. They have high conductivity, making it easy for them to pass *current*, a flow of charges or electrons.

There are millions of free electrons within solid metal conductors of electricity that can move or wander among the atoms. In other words, these electrons can be temporarily detached from the atoms. When a conductor connects a charged body to ground, the free electrons flow in a continuous stream. They transfer the charge by flowing in a definite direction. Because electrons have the same negative charges, this movement is a result of the force of *repulsion* between the electrons and the force of *attraction* between the electrons and the positive charges.

There are very few free electrons in insulating materials such as rubber, glass, and mica; consequently they oppose the flow of electric currents. Because of this opposition, these materials have very low conductivity and so are classified as *nonconductors*, or *insulators*.

The opposition to the flow of electrons is called *resistance* and it is measured in ohms. The resistance, R, of a material is inversely proportional to its cross-sectional area, A, perpendicular to the flow of current and is directly proportional to the length of the material parallel to the flow of the current. Consequently, the thicker a conductor, the lower its resistance, but the longer the conductor, the greater its resistance. This can be expressed as:

$$R = \varrho_v \frac{l}{A}$$

where: ϱ_v = a constant known as volume resistivity.

Volume Resistivity

Volume resistivity is expressed in ohms per centimeter and is a constant for a given homogeneous material. The equation can be rewritten as:

Conductors and Insulators

$$\varrho_v = \frac{RA}{l}$$

Thus, the volume resistivity of a homogeneous material can be determined by measuring the resistance of a piece of the material with known dimensions; that is, length, l; width, w; and thickness, t.

For a square piece of material in which l is equal to w, the equation can be simplified to:

$$\varrho_v = Rt$$

Thus, ϱ_v is normally determined by measuring the resistance, R, of a square of material and multiplying it by the thickness, t.

Volume resistivity is an inverse measure of the conductivity of a material. In an electrical insulating material, it is numerically equal to the volume resistance in ohms between opposite faces of a 1 cm cube of the material.

Volume resistivity is the primary unit of measurement, but surface resistivity, or *sheet resistance*, is also frequently used. Originally it was a convenient term applied to thin films of materials, generally metallization layers. It is now used by industry to describe the conductivity of much thicker materials.

Surface Resistivity

Surface resistivity is numerically equal to the surface resistance, R, of a square section of material of a given thickness. (The size of the square is irrelevant.) It is measured in ohms per square.

Surface resistivity is commonly used as a resistance measurement variable for laminated materials with thin conductive surfaces over an insulative base. It is used to define the resistivity of surface-conductive materials such as hygroscopic antistatic plastics such as polyethylenes, and other conductively coated or laminated insulative materials.

Conductive layers on these materials, such as the sweat layers of hygroscopic antistatic materials, usually have near uniform thickness. The surface resistivity does not change significantly by increasing or decreasing the thickness of the base insulating material if its volume resistivity is high in relation to that of the conductive surface materials. In military specifications, surface resistivity measurements are used to describe three basic types of materials used for ESD protection: conductive, static dissipative, and antistatic.

Conductive materials have surface resistivities of 10^5 ohms per square or less. Metals, some bulk conductive plastics, wire-impregnated materials, and laminates meet this requirement. A bulk conductive material with a volume resistivity of 10^4 ohms per centimeter would be conductive if its thickness were 0.1 cm or greater. However, if the material had a thickness of less than 0.1 cm, it would be classified as static dissipative.

Static-dissipative materials have surface resistivities of greater than 10^5 but less than 10^9 ohms per square. Some materials normally considered to be conductive can be formed into films so thin that they will have a surface resistivity in the static-dissipative region, and so will be classified as static dissipative.

Antistatic materials have surface resistivities equal to or greater than 10^9 but less than 10^{14} ohms per square. (Some standards and specifications use 10^{13} as the upper limit.) These materials include hygroscopic antistatic materials such as some melamine laminates, high-resistance bulk conductive plastics, wood and paper products, and very thin layers or films of static-dissipative or conductive materials.

Hygroscopic (water-seeking) agents are added to many plastic antistatic materials during manufacture. These agents constantly migrate to the surface, where they attract atmospheric moisture. The resulting monolayer of water forms a thin, electrically conducting surface that dissipates static charges.

If conductive carbon powder is added to a liquidifed plastic vehicle, it can produce a homogeneous electrically conductive material. To form an effective conductive path to ground, it might be necessary to add up to 40 percent carbon by volume to the plastic vehicle. However, this high percentage of a nonplastic ingredient could weaken the plastic structurally, allowing it to tear and puncture more easily.

Another method for lowering the resistivity of inherently insulating plastics is to add metallic fibers, forming *metalloplastics*. Resistivities as low as 0.001 ohm per centimeter have been obtained commercially. In another process for manufacturing static dissipative materials, only 10 percent of the material is a conductive chemical additive. The additive is reported to bond chemically with the molecular structure of the base polymer to produce an ionic path through the entire volume of the material. These materials are said to be permanently antistatic and not dependent on moisture for conductivity.

CHARGE DISSIPATION

Once a charge is generated, its distribution depends on the resisitivity and surface area of the material. In other words, the more conductive the material, the faster the charge is distributed. The greater the surface

Charge Dissipation

area over which a charge is spread, the lower the charge density and the level of the residual voltage. As stated earlier, localized charges cannot exist on conductors.

Because of this effect, conductive objects and materials are used for electrostatic discharge control. However, there are distinct limitations for these materials. If, for example, a charged circuit board or semiconductor device is brought near a highly conductive object such as a tote box or tabletop, a spark and high discharge current could occur. When this happens, some semiconductor devices in the discharge path could be damaged. To prevent this occurrence, all boxes or trays for carrying ESD-sensitive devices, as well as all work surfaces and flooring in ESD-controlled areas, should be made from materials conductive enough so that significant voltages will not be induced across the surfaces, but not be so conductive that a spark discharge will occur.

When an object with an electrostatic charge is placed on a static-dissipative or conductive surface, the charge will gradually dissipate or decay. Although this decay may appear to be instantaneous, its duration may be several hundredths of a second up to several seconds.

Decay time is generally measured by charging a section of material with a static voltage and measuring the time for the voltage to decay to a specified level, such as 10 percent of its original value. The Electronic Industries Association (EIA), for example, specifies a decay time for bags and pouches used in carrying ESD-sensitive components. The time taken for either an applied positive or negative 5,000V charge to dissipate to 50V after the material is grounded should not exceed 2.0 seconds.

The Department of Defense (DOD) and the National Fire Protection Association (NFPA) also have prepared specifications for static decay testing. In general, decay time is related to conductivity.

An electric current is considered to be a flow of electrons. However, in gases and liquids the flow of electricity is a movement of ions, rather than of electrons. If an atom or molecule loses one or more electrons, it becomes a positive ion; if it retains more than its normal share of electrons, it becomes a negative ion.

Under ordinary circumstances, air is a poor conductor because most of its molecules are electrically neutral. Some of the molecules will be ionized by ultraviolet light or cosmic rays. When two strong opposite charges are brought close together, the force between them may be enough to impart sufficient energy to these random ions so they collide with other molecules. In so doing, they dislodge electrons and create additional ions. In a brief instant, a chain reaction occurs, making the air a conductor and allowing a sudden rush or discharge of electrons to bridge

Principles of Electrostatics

the gap between the two molecules. The resulting arc is an electric spark.

The transfer of charge by ions can also take place in liquids. Water can be chemically, but not necessarily biologically, pure if it contains no dissolved minerals. Pure water is, however, a very poor conductor. It is usually obtained by distillation in the laboratory because even rainwater contains dissolved minerals. Safe drinking, or *potable*, water is a fair conductor because of its dissolved minerals.

It is difficult to perform classroom and laboratory demonstrations of static and its properties in rooms where the relative humidity is high. On damp days, moisture condenses on many objects. Although not necessarily visible, it forms a very thin coat or film of water, allowing charges to dissipate.

Movement of ions in films of water and air may be a better explanation of some triboelectric charging than is transfer of electrons. In the classic experiment of rubbing an amber rod with fur, it has been assumed that these objects become charged because electrons have been moved from one object to the other. Because amber and fur are poor conductors, however, they would not be expected to have any significant number of free electrons. The charge generated could be due to the movement of ions.

The earth is another good conductor unless the soil is extremely dry and/or rocky. In some desert areas of the world, the ground is so dry that it must be soaked with water to form an effective electrical ground.

If a charged object touches or is connected to ground, it is said to be *grounded*, or *earthed*. It is not always necessary for the ground connection to be made by a metallic conductor. The human body, for example, can act as a ground conductor, although it is not very effective.

Human body resistance can vary from 100 to 100,000 ohms, depending on the amount of moisture, salt, and oils on the surface of the skin, skin contact area, and pressure. Typical values of 1,000 and 5,000 ohms are measured for hand-holding, hand-shaking, and the grasping of objects. A value of 1,500 ohms has been selected to represent the human body in DOD standard ESD simulation circuits as discussed later in this chapter and in chapter 9.

Because the human body can have a low resistance, common voltages present in some assembly and test procedures can be lethal if a person accidentally touches an electrically live circuit and ground at the same time. To minimize these risks, it is common practice in ESD-controlled workstations to employ *soft grounds*. A soft ground is a connection to ground through a resistor whose value is high enough to limit current flow through the human body to a safe level, normally 5 milliamperes. Under normal conditions this resistor value is 1 megohm; however, the value of that resistor should be increased to take into account any voltages present on exposed terminals within reach of a grounded person at a workstation.

Charging by Induction

A typical example would be the terminals of a benchtop power supply or active voltages in a unit of equipment under test.

The 1-megohm, soft ground resistor is most commonly used in series with a personal wrist ground strap and the ground connection. If no resistor is present and the wrist strap is grounded directly, it is said to be a *hard ground*.

CHARGING BY INDUCTION

Most conductors are charged by *conduction*, when there is physical contact between the charging source and the object being charged. Charging also can be done by *induction*, when there is no physical contact between the charging source and the object being charged. (See FIG. 2-4.)

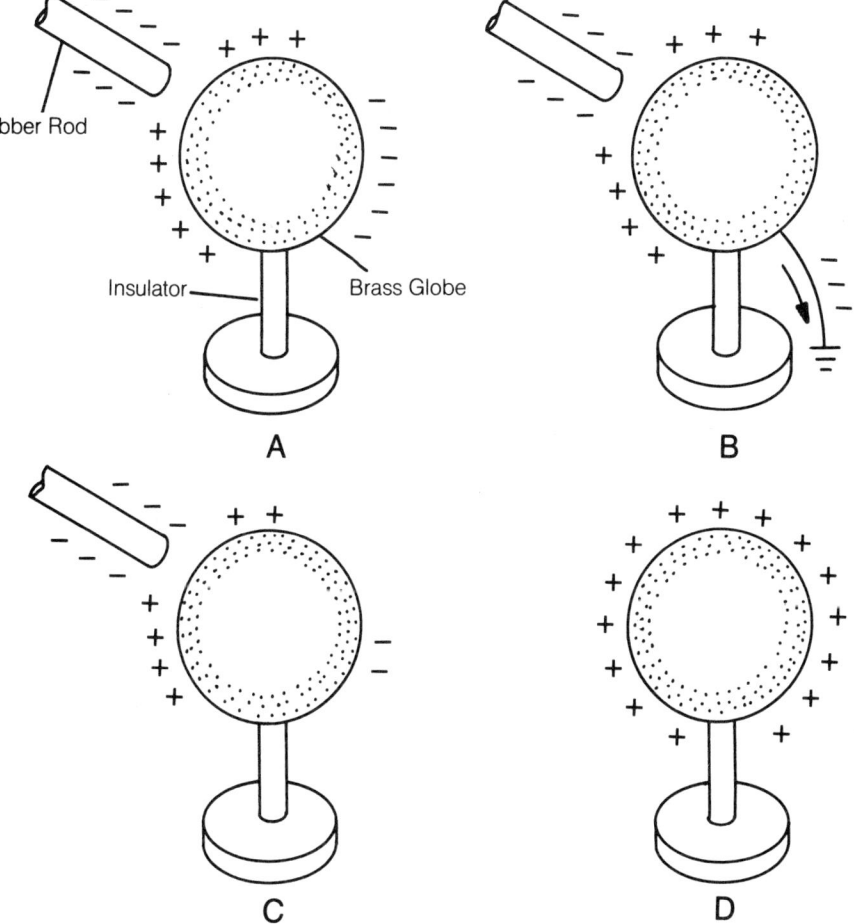

Fig. 2-4. Charging a metal globe by induction.

Principles of Electrostatics

When a negatively charged hard rubber rod, for example, is brought near an insulated metal conductor, as shown in A, the electrons in the rod will force many of the electrons in the conductor to the opposite side of that conductor. The conductor's shape is not relevant to this phenomenon; it can be a sphere, rod, plate, or other shape.

While the rod is still in the vicinity of the conductor, a person can touch the conductor or connect a wire from it to the ground, as shown in B. The electrons in the conductor flow through the person or the wire to the ground as they are being forced away from the rod. After a brief interval, the ground connection is broken, as at C. With no free electrons left on the conductor, the conductor now has a net positive charge. The positive charges distribute themselves evenly over the conductor, as shown at D, when the rod is removed from the vicinity.

If a positively charged rod is used, it would attract negative charges (free electrons) to the near end of the conductor, leaving the far end with a positive charge. By grounding the conductor, electrons are pulled up into the conductor from the earth, giving the conductor a net negative charge.

Whenever a charged object is brought near a conductor insulated from ground, it causes charge separation, leaving a charge with the same polarity, or sign, at the opposite end or side of the conductor and a charge of opposite sign at the near side of the conductor. In the process, the charging object does not lose any of its charge. The magnitude of the charge that is induced on an object depends on the object's size, shape, and proximity to the charging body.

An investigation of IC shipping magazines, commonly called *DIP sticks*, showed that charges were being developed on the ICs because of their movement in the tubes. (The term *DIP* is derived from dual-in-line package, an IC package style in which pins project from the long, parallel sides of the insulating body.) A mobile charge is formed on the metal leads and conductors of the DIP-packaged IC and an immobile charge is formed on the epoxy-plastic case of the IC, as shown in FIG. 2-5. Although the immobile charge cannot develop currents (and therefore cannot directly damage the IC), it can produce a mobile charge by induction and electrostatic discharge on the leads of the IC. This situation occurs when the immobile charge induces a charge separation on the leads. Then, if the leads are grounded, a discharge takes place.

ELECTRIC FIELDS

Earlier it was stated that electrically charged bodies can attract or repel each other through a distance although they are not linked by any solid, gas, or liquid. This response can occur in a vacuum. There is no precise explanation of how this action occurs at a distance. Scientists have

Electric Fields

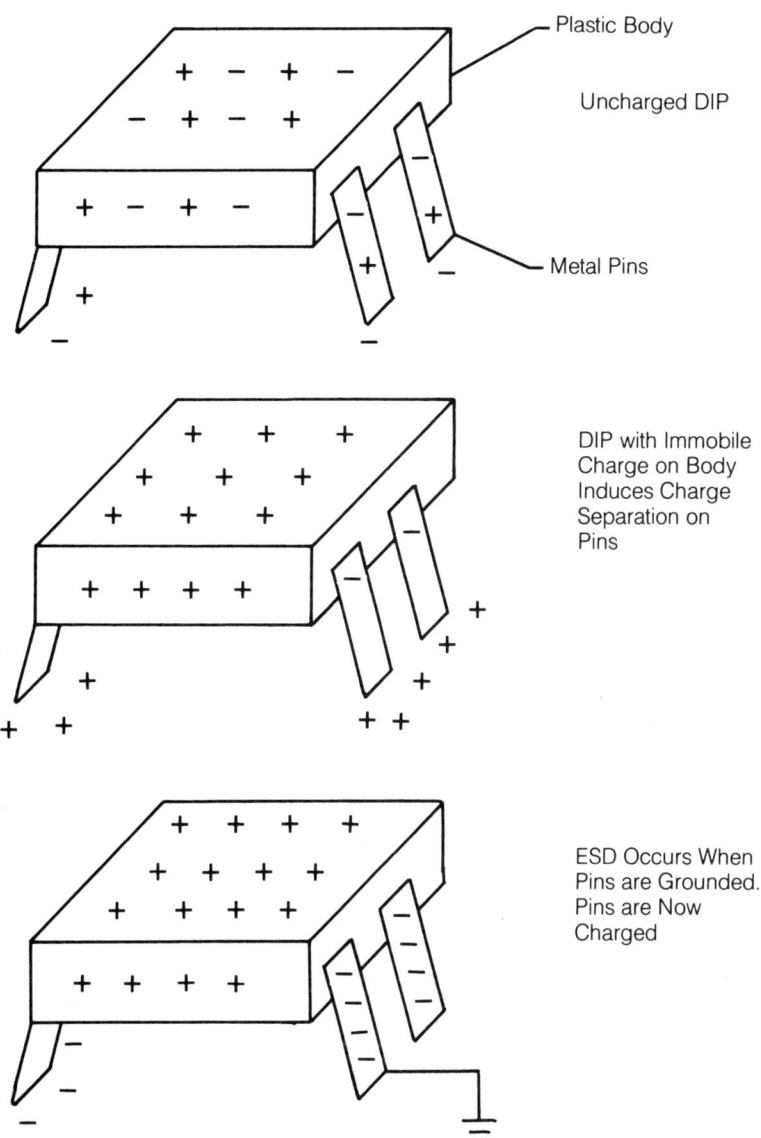

Fig. 2-5. Charge induced on dual-in-line (DIP) package leads as a result of immobile charges on insulated package body.

devised a model to make this concept easier to understand. It has two distinguishing features: a force field and lines of force.

The space or region in the vicinity of an electric charge is not the same as an identical space with no charge in it. An object placed in this space has different properties (force, charge, potential) than it would have

31

Principles of Electrostatics

in a space with no nearby charge. The charge is said to alter the space, although there is nothing tangible there to be changed, at least in a vacuum. The altered space is called an *electric field*.

At any point in this field, an electric charge, q, will encounter either a force of attraction or a force of repulsion. The force, F, is proportional to the charge, q, and the strength, E, or intensity of the electric field. That is,

$$F = qE$$

where: E and F are vectors.

If q is positive, the force will be in the direction of the field. If q is negative, the force will be opposite the direction of the field.

The electric field strength, E, at any point is the force, F, on a small positive test charge, q_1, divided by q. That is,

$$E = \frac{F}{q_1}$$

In this calculation it is assumed that the test charge is not large enough to upset the field or change the charge distribution to give a false indication. For most practical purposes, the electric field has influence in the space immediately adjacent to the charge that created the field. In theory, however, the field extends to infinity. Nevertheless, its strength is negligible at distances beyond the immediate vicinity.

The equation $E = F / q$ can be combined with Coulomb's Law to show that the electric field strength at a point a distance, r, from the charge is:

$$E = K \frac{q}{r^2}$$

where: k = a constant.

There is no electric field inside a solid conductor if there is no current through the conductor. If there were an electric field in this conductor, it would cause the free electrons to move, but electrostatics is the science of charges at rest.

The English scientist Michael Faraday demonstrated that an electric field cannot exist inside a closed metal container as shown in FIG. 2-6. A Faraday cage (named after him) can demonstrate this principle. The cage

Electric Fields

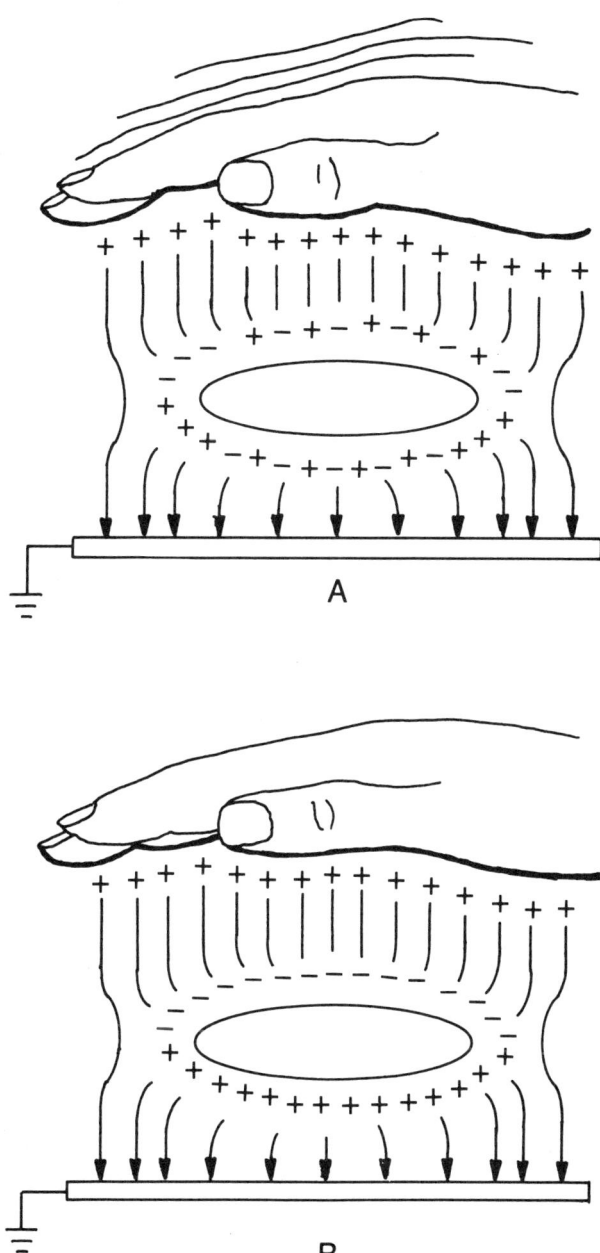

Fig. 2-6. No electric field exists inside a Faraday cage and the charge at A is redistributed as a charged body approaches as at B.

Principles of Electrostatics

is widely used for electrostatic shielding because an enclosed object is protected from external fields and discharges. If there is no electric field within a conductor, all of the charge on the conductor must lie on its surface, as shown in the figure. Charge is redistributed around the closed metal container when the charged object, a hand, is brought near it.

If instead of an isolated charge, a number of individual charges are in the region, the electric field becomes complex. Its intensity at any given point will be the vector sum of the electric field from all these charges.

The concept of an electric field can be shown using lines to represent imaginary lines of force giving the direction of electrical intensity. As shown in FIG. 2-7, these can be straight lines or curves. The arrow on each line indicates the direction a small positive charge would take if it were placed at the point. (A negative charge would travel in the opposite direction.)

Electric field lines are three dimensional. The lines from a point charge, for example, project out like quills on a porcupine. Because of limitations of an isometric drawing, it is not obvious that these lines are continuous between positive and negative charges. So, each line must start on a positive charge and end on a negative charge forming a continuous bridge. For isolated charges, the opposite charge is imaged at infinity; that is, each line extends to infinity as shown in FIG. 2-7A.

For an isolated negative charge, the arrows would point in the opposite direction from that shown in FIG. 2-7A, indicating that the lines started at infinity. To show variations in field strength originating or terminating on an object, the lines are drawn at different densities. However, no matter how closely they are drawn in this convention, the lines never cross.

In FIG. 2-7B, the strongest lines of the electrostatic field between positive and negative charges form the contour of a football or baker's whisk. In FIG. 2-7C, the charged lines repel each other creating opposing convolutions. The field lines, as illustrated in FIG. 2-7D, are those of a flat-plate capacitor.

ELECTROSTATIC MEASUREMENTS: THE ELECTROSCOPE

Although many different kinds of electroscopes are used to detect static charges, the gold-leaf version invented about 200 years ago is the landmark configuration. Figure 2-8 is a cutaway view of this classical instrument. It consists of a conductive plate, rod, and leaf assembly mounted inside a grounded metal case with an insulating sleeve. The metal shaft has a circular metal plate on top and two thin gold strips called *leaves* fastened on its lower end. The electroscope has a glass window for observing the deflection of the leaves.

Electrostatic Measurements: The Electroscope

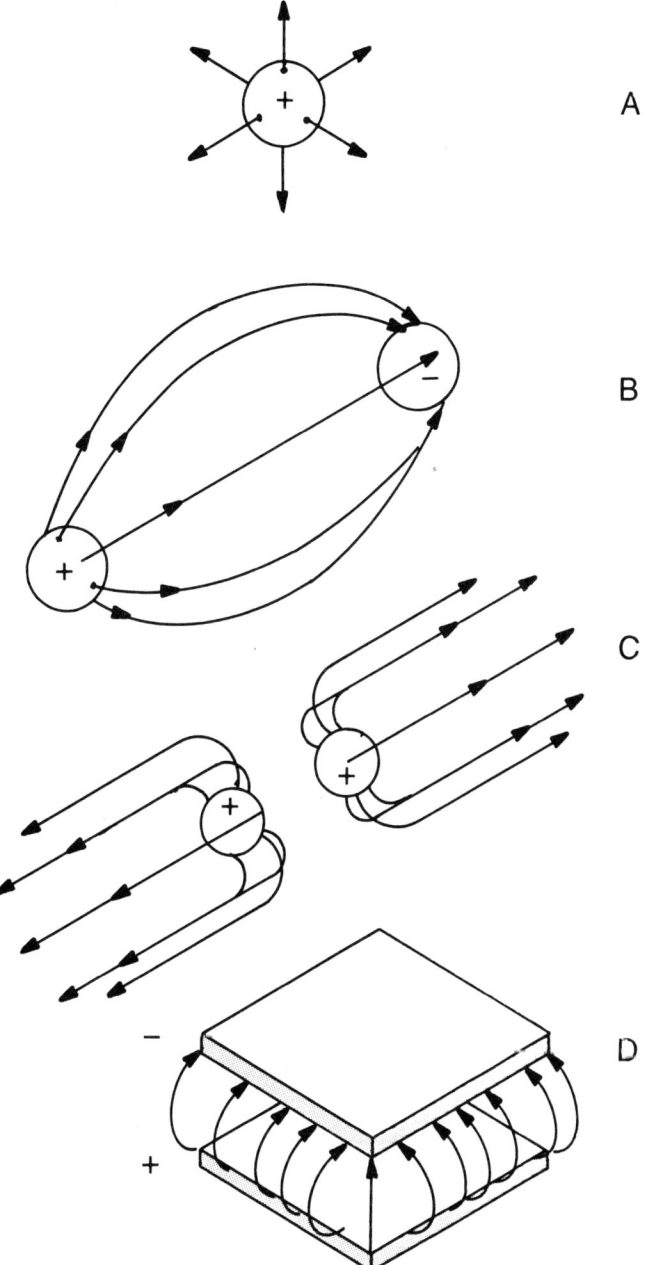

Fig. 2-7. Electric fields: an isolated charge (A), unlike charges (an electric dipole) (B), like charges (C), and parallel flat plates (D).

Principles of Electrostatics

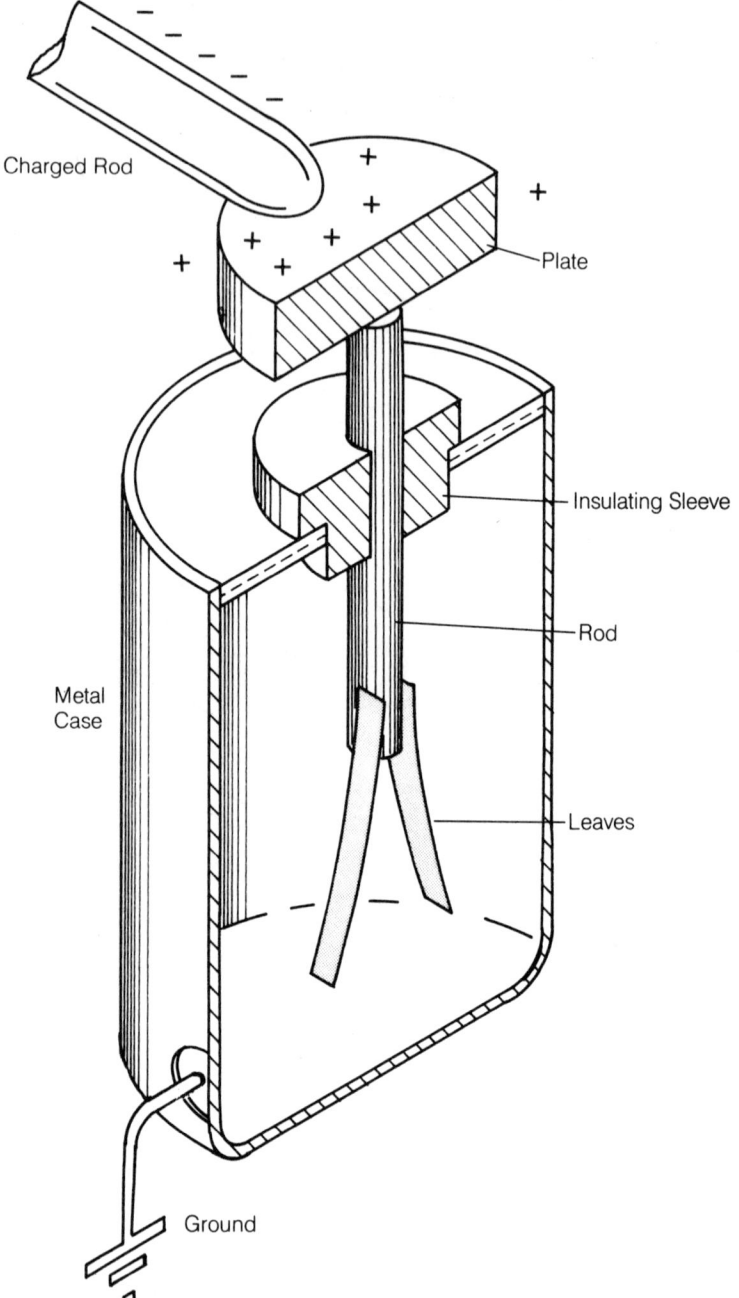

Fig. 2-8. Gold-leaf electroscope.

Electrostatic Measurements: The Electroscope

This form of electroscope also permits a quantitative measure of electrostatic charge. Because the metal leaves are sealed in the metal and glass case, they cannot be disturbed by air currents in the room. Experiments have shown that the angle of separation between the leaves is proportional to the charge. Modern versions of this classic instrument usually use aluminum foil, rather than gold leaves.

When the metal plate on top of the electroscope touches a positively charged body, as shown in FIG. 2-9, the body will attract electrons from the leaves, rod, and plate. This gives the leaves a positive charge. Because the leaves have like charges, they will fly apart as a result of repulsion. The greater the charge, the farther the leaves will separate.

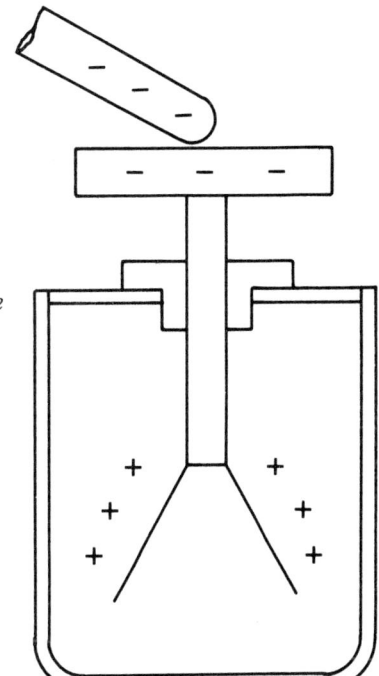

Fig. 2-9. Charging an electroscope with a positive source.

This example is shown with a positively charged body, but the electroscope works equally well with negatively charged bodies, as shown in FIG. 2-10. Whether positive or negative, the charge to be detected or measured immediately distributes itself over the leaves. As a result, the leaves will have the same sign (positive or negative) as the charging object.

37

Principles of Electrostatics

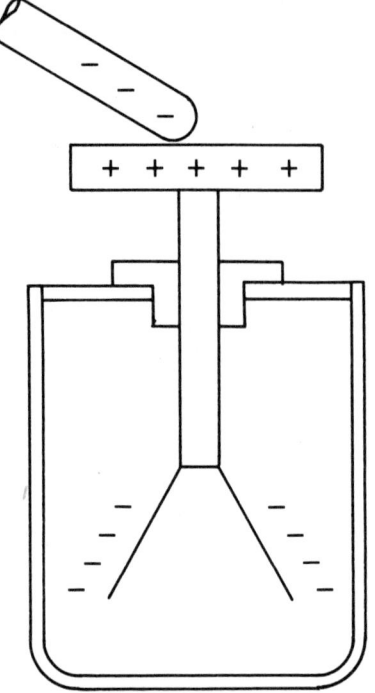

Fig. 2-10. Charging an electroscope with a negative source.

If the charge on the electroscope is known, the electroscope can be used to determine the sign of an unknown charge. For example, if the electroscope is positively charged, as shown in FIG. 2-11A, and a charged body touches the electroscope, its polarity can be determined by the reaction of the leaves: if they diverge further, the rod has a positive charge; if they converge, the rod has a negative charge as shown in FIG. 2-11B.

An electroscope can be charged without direct contact. (This may be done if there is reason to believe that the unknown charge is so great it could damage the electroscope.) To charge the electroscope by induction, the electroscope is first grounded (FIG. 2-12A), then the charged object or rod is held near the plate. Inductive charge separation is shown in FIG. 2-12B. The electroscope leaves are inductively charged with a charge opposite to that of the source charge, causing the leaves to deflect. As the charged rod is removed, the leaves collapse again. However, if the plate of the electroscope is touched by the hand or is grounded by a wire before the charged body is removed, the leaves collapse even when the charged body remains nearby.

As in the case of charging by direct contact, when a charge of the same sign is brought near the plate, the charge increases the leaf deflec-

Electrostatic Measurements: The Electroscope

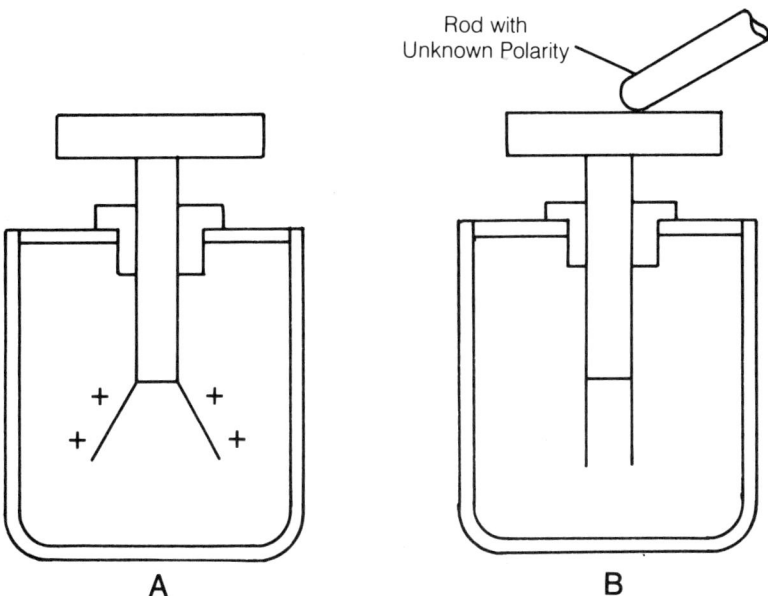

Fig. 2-11. Using an electroscope to determine the polarity of a body. Electroscope with positive charge (A), and after being touched by a body of unknown polarity (negative to drain the positive charge B).

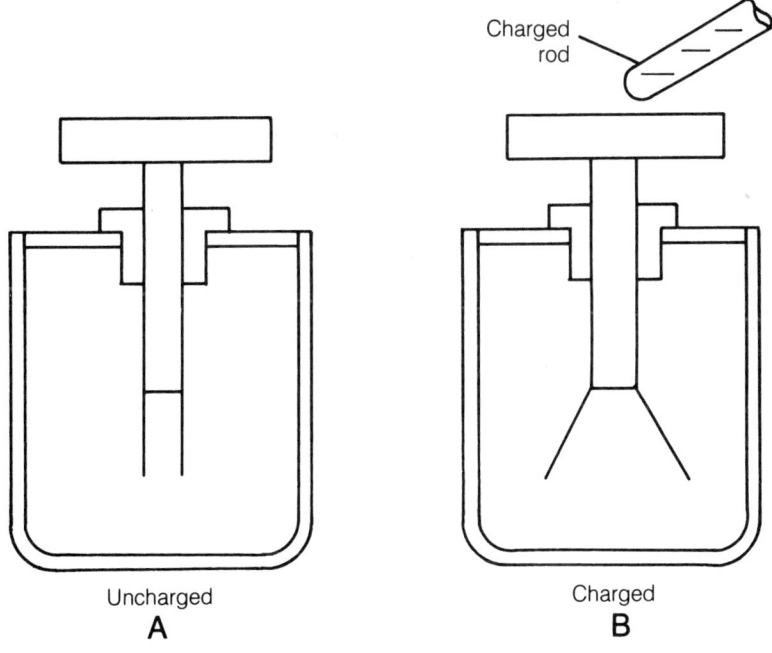

Fig. 2-12. Charging an electroscope by induction.

Principles of Electrostatics

tion. For an opposite external charge, the charge will be drawn from the leaves and rod, decreasing leaf deflection. The angle of leaf deflection indicates the potential difference between the electrically isolated plate leaf assembly, and metal case, and is proportional to the potential difference.

Because the leaves have like charges, they will fly apart. The greater the charge on these leaves, the farther they will separate, increasing the angle of divergence.

To convert an electroscope from a charge-measuring instrument to a potential-measuring instrument (i.e., a simple voltmeter), a scale can be added to measure the angle of deflection of the leaves. To make a measurement, the plate on the electroscope is connected by a wire to the body whose potential is to be measured.

ELECTRICAL POTENTIAL

In FIG. 2-13, a small positive test charge, q, is placed in an electric field associated with charge, Q. Work will be required to push or move it from point A to point B because energy must be expended to overcome the force of repulsion. This work is stored as potential energy in q. If the test charge is moved from point B to A, the electric field will do work and the potential energy of q will decrease. The work, W, done in moving the positive charge is the *potential difference*, V, between the two points.

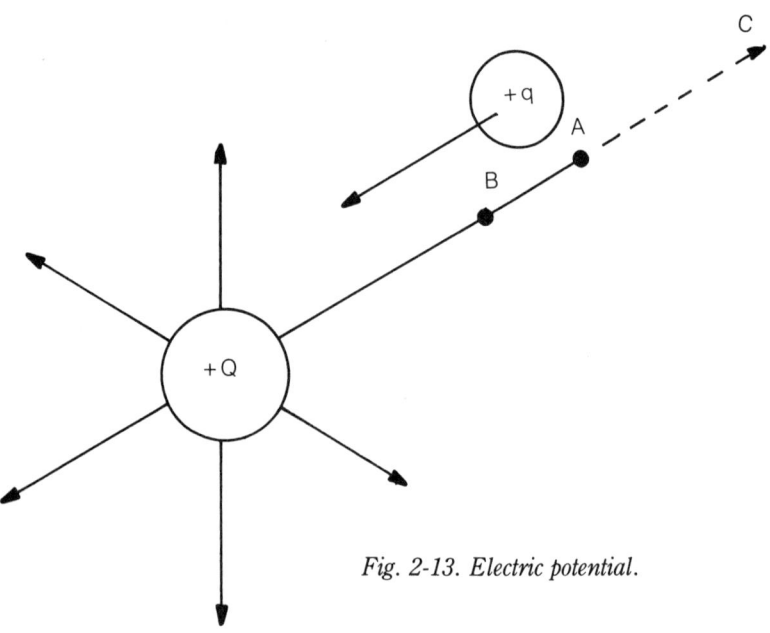

Fig. 2-13. Electric potential.

Electric Potential

The three quantities are related by the equation:

$$\text{Potential difference} = \frac{\text{Work done}}{\text{Quantity of charge transferred}}$$

Or,

$$V = \frac{W}{q}$$

When q is in coulombs and W is in joules, V is expressed in volts. Thus, if 1 joule of work is required to move 1 coulomb of charge from point A to point B, the potential difference between the two points is 1 volt.

Because potential difference is a scalar quantity, it can be added to other potential differences algebraically. The potential difference between two points is also referred to as the *voltage* between the two points.

In FIG. 2-13, the path of q from A to B is a straight line. However, even if the path had been curved and distorted, the work required would have been the same. A positive charge will tend to move from a point of high potential to a point of lower potential (from B to A in FIG. 2-13). Point B will be at a higher potential than point A because it is closer to the source.

The line of force always points from higher to lower potentials for positive charges. Point B may have, for example, a potential of 120 volts as referenced to earth, and point A may have a potential of 100 volts. The potential difference between A and B is simply the difference of 20 volts. Work is required to move a positive charge from a point of lower potential to one of higher potential; the converse is true for negative charges.

In making measurements of difference of potential, it is useful to determine the voltage between a point and the earth (ground). The potential of the earth is considered to be zero. Another point considered to be zero is infinity.

Because of the earth's zero potential, any conductor connected to it will discharge and also have a zero potential. Similarly, any two conductors will have the same potential when connected together once the circuit has settled down and current no longer flows. There is no electric field inside a hollow, closed conductor, but it can contain a region of uniform potential, which will be the same on the inside wall as on the outside wall.

EQUIPOTENTIAL SURFACES

Within an electric field are many points with the same potential. If imaginary lines are drawn connecting these points, an *equipotential surface* is represented. Since all points on this surface have the same potential, no work is required to move a charge around on it.

Principles of Electrostatics

When the surface is drawn, it will be at right angles to the electric field. The surface may be a plane but generally it has some curvature. In the case of an isolated point charge, the equipotential surfaces will form concentric spheres around the charge, as shown in FIG. 2-14. Only two spheres are shown, but an infinite number could be drawn.

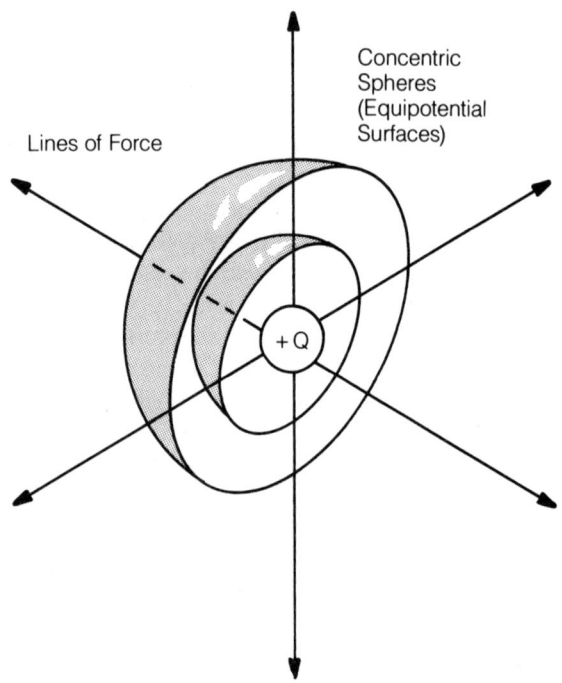

Fig. 2-14. Equipotential surface.

Figure 2-15 shows the electric force on charge q moving at right angles to the line of force. No work is required to move q from A to C, a short distance. If q keeps moving at right angles to each line of force it encounters, it will map out an equipotential surface. Although no work is done in moving a charge on an equipotential surface, work must be done to move a charge from one equipotential surface to another, as from the outer sphere in FIG. 2-14 to the inner sphere.

As stated earlier, a static electric field cannot exist within a conductor when the charges in the conductor are at rest. In electrostatics, the consequence is that the surface of a single conducting body becomes an equipotential surface; there will be no difference of potential between any two points of a conductor. Here again the electric field surface will be perpendicular to the conductor.

Surface Charge Density

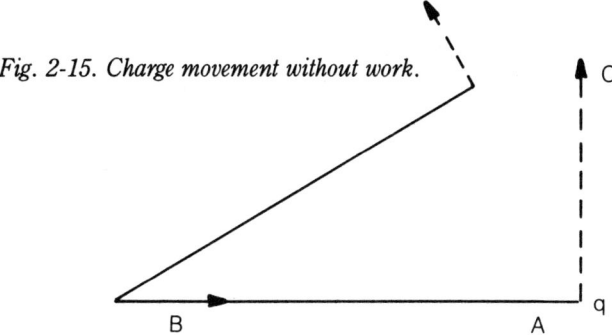

Fig. 2-15. Charge movement without work.

SURFACE CHARGE DENSITY

Electrostatic charges exist only on the outside surface of a conductor, whether it is solid or hollow. The charges are distributed uniformly over spherical conductors; however, the charges are not distributed evenly over conductors with nonuniform shapes, such as footballs or eggs. Figure 2-16 shows that the charges are greatest where the surface has the most curvature, in this case at the ends or projections. The sharper the curve, the greater will be the concentration of charge or charge density.

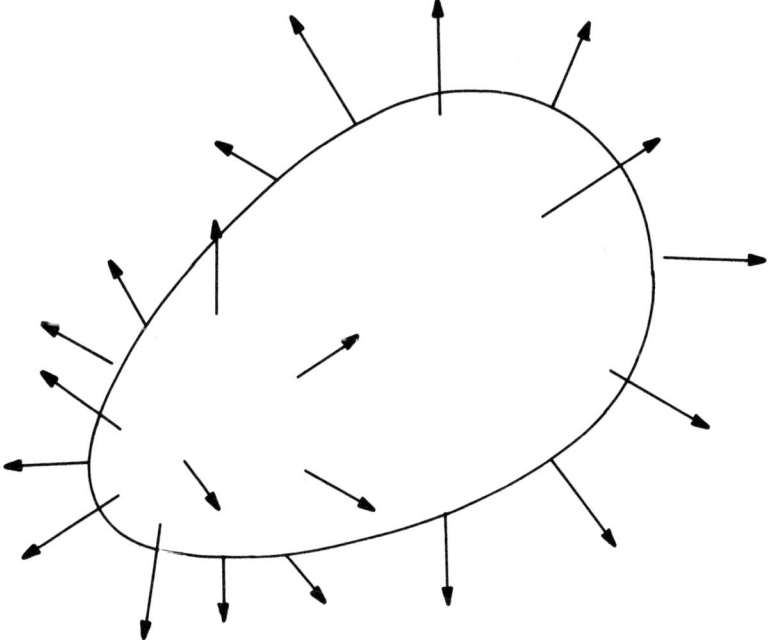

Fig. 2-16. Surface charge density.

Principles of Electrostatics

The chances for the charge to escape are greater at curves with a small radius than those with a large radius, and they are greater at points than on any curved surface. Charges escape as the surrounding air becomes ionized, initiating corona discharge.

Corona discharge is conspicuous from pointed surfaces shaped as spikes or needles. Benjamin Franklin discovered this principle and used it in the design of the lightning rod. In general, a grounded sharp metallic point can discharge conductors without sparking simply by bringing the point near the conductor.

Sharp or pointed surfaces should be avoided anywhere an electric charge is to be retained. Instead, spherical surfaces with a large radius of curvature should be used. The larger the diameter of the sphere, the greater its potential will be before the surrounding air breaks down and conducts. This principle is employed in the Van de Graaff generator discussed later in this chapter.

CAPACITANCE

Two conductors held close together without touching, separated by a dielectric, form a *capacitor*. At low frequencies, the conductors will normally be parallel metal plates, as shown in FIG. 2-17A. A capacitor can store electric charges when its plates are connected to a voltage source such as an alternating current (ac) generator or battery. When connected in this way, one plate will acquire a positive charge and the other a negative charge. The charge, Q, of the capacitor is considered to be that of either conductor because the charges are equal but opposite.

Because the charges are separated by a *dielectric* or insulator, such as air or mica, they cannot flow through the capacitor. Instead, they accumulate on the plates and remain until the potential is removed, the polarity is reversed, or a shorting wire is connected between the two conductors. If the potential is simply removed from the capacitor, the charges eventually leak away. If the polarity is reversed, however, positive and negative charges on the plate effectively change places. If a wire is connected between the plates, the energy is released suddenly and could spark under certain circumstances.

Charge Storage

The ability of a capacitor to store a charge is called *capacitance*, C. It is directly proportional to the charge, Q, on either conductor and inversely proportional to the voltage (potential difference) between the conductors. The relationship is:

$$C = Q / V$$

where: C = farads
Q = coulombs
V = volts

The unit of capacitance, the farad, was named in honor of the English scientist Michael Faraday. Because it is a large unit, the fractional units of the *farad*—the microfarad (μF) and the picofarad (pF)—are more commonly used. In modern electronic equipment, capacitors have values that typically range from 1 pF (10^{-12} farad) to thousands of microfarads (10^{-6} farad).

Energy Stored in a Capacitor

When a capacitor is charged, work must be done to transfer the charges from one plate to the other. The total work, W, in this process is related to the charge and the voltage by the expression:

$$W = \tfrac{1}{2} QV$$

This may also be written as:

$$W = \tfrac{1}{2} CV^2$$

Determination of Capacitance

The capacitance value of a capacitor depends on the distance between the plates, d (FIG. 2-17B), the nature of the dielectric separating the plates, K, and the size of the plates, A. Capacitance is directly proportional to plate size and dielectric constant, and inversely proportional to plate separation. This relationship is expressed mathematically as:

$$C = K\frac{A}{d}$$

where: A = area plates in square meters
d = distance between plates in meters
K = dielectric constant

Dielectrics

When a solid dielectric is inserted between the plates of a charged capacitor separated by air, the electric intensity will weaken, and the potential difference, V between the plates will decrease. In effect, for a

Principles of Electrostatics

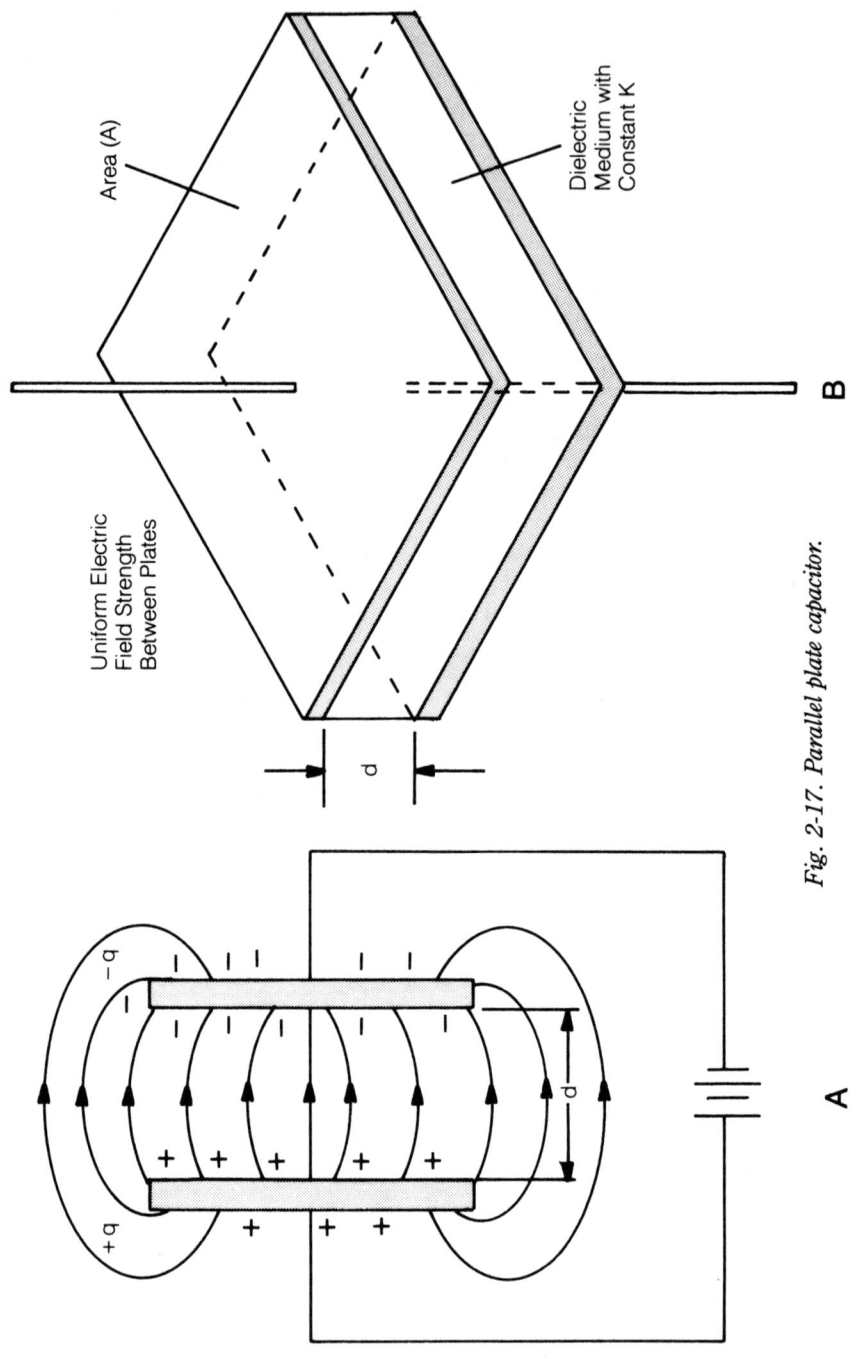

Fig. 2-17. Parallel plate capacitor.

Capacitance

given fixed charge, when the potential is decreased, the capacitance will be increased, as can be seen from the equation:

$$C = Q / V$$

The increase in capacitance will depend directly on the type of material used as a dielectric. This factor, the *dielectric constant*, is a simple number. The higher this constant, the greater the capacitance will be. Representative values are 1 for air, 1 to 3 for many common plastics, 4.5 to 7.5 for mica, and 4.8 to 10 for glass and ceramics.

Electrons in a nonconductor cannot leave the atom or molecule in which they are located. In the absence of an electric field, the molecules arrange themselves as shown in FIG. 2-18A. However, in the presence of a positive electric field, they may shift their positions. When this shift occurs, the molecule becomes an electric dipole and the material is *polarized*. These molecules line up as shown in FIG. 2-18B. Within the interior of the material, the plus and minus charges cancel each other, leaving the material neutral.

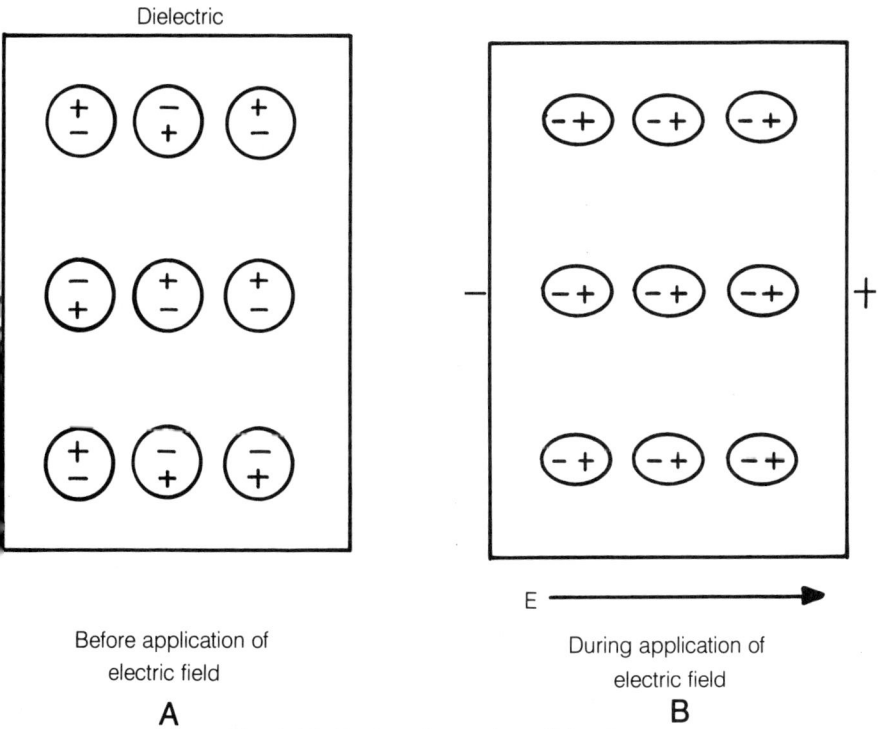

Fig. 2-18. Surface charges in a dielectric.

Principles of Electrostatics

In this case, the dipoles on each side of the material are affected. In the absence of a matching charge, the left side will have a negative charge and the right side a positive charge. These charges are called *bound charges*. They reduce the electric field intensity because they are in opposition to it.

Capacitors as Components

Capacitors for laboratory experiments may be made as concentric spheres, concentric (coaxial) cylinders, or parallel plates. However, most commercial electronic capacitors for use as electronic components are variations of the parallel plate capacitor shown in FIG. 2-17B. Made in a bewildering range of sizes, shapes, capacitance values and materials, the basic parallel plate design is not always easy to recognize.

The variable air-dielectric tuning capacitor for radio receivers has interleaved aluminum plates on a rotating shaft that mesh with other fixed plates on the same shaft. By turning a control knob, the position of the moving plates may be changed with respect to the fixed plates. This motion serves to alter the frequency of the tuning circuit in which the variable capacitor is located. The parallel plate design is not as easy to recognize in fixed-packaged capacitors with ceramic, glass, and mica dielectrics.

The manufacture of fixed capacitors has become an international business valued at billions of dollars per year. Commercial fixed capacitors are classified as *electrostatic* or *electrolytic*. They are packaged in a wide variety of forms with radial or axial leads, or leadless for surface mounting. Capacitors can be dipped or molded in protective plastic or encased in metal jackets.

The monolithic multilayer ceramic capacitor (MLC), illustrated in the cutaway drawing of FIG. 2-19, is a variation on the basic parallel plate capacitor. Metal "inks" are deposited on thin sheets of "green" ceramic to form the electrodes, or " plates." The raw ceramic sheets are then compressed, fired, and diced to form monolithic capacitors in a wide range of sizes, performance characteristics, and values.

The end surfaces are metallized forming electrical bonds between alternate internal plates; the end terminations are formed by additional metallization and plating. Unpackaged monolithic ceramic capacitors are bonded directly to metallized pads on hybrid microcircuit ceramic substrates or metallized pads on surface-mount circuit boards. The leads of coated and leaded MLCs are inserted in the plated-through holes of conventional circuit boards.

Paper and film-type capacitors are cylindrical, but are basically sandwiches of metal foil with paper or plastic film dielectrics rolled around axial leads. A variation of this design is the metallized-film capacitor. Even foil-

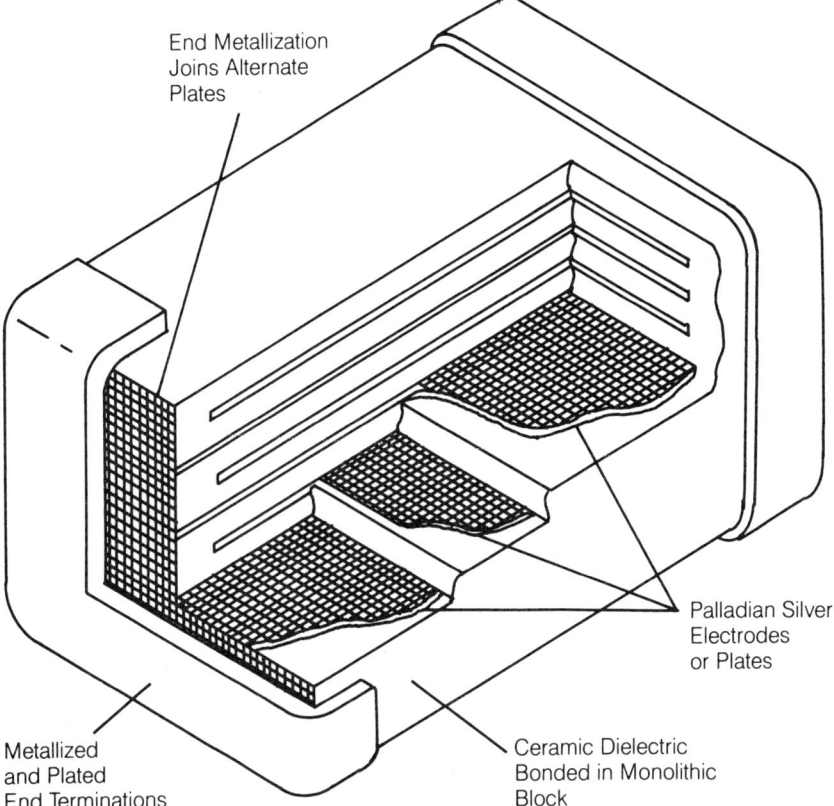

Fig. 2-19. Ceramic monolithic multilayer capacitor chip is a form of parallel plate capacitor.

type tantalum and aluminum electrolytic capacitors are fundamentally parallel plate capacitors. The anodes and cathodes are separated by paper before they are rolled and inserted in cylindrical metal cases.

Unintentional Capacitors

The capacitors discussed so far are discrete units used as circuit components. However, a capacitor is formed by the presence of two neighboring conductors. These include conductive paths on circuit boards and ground wires with air dielectric between them. These virtual capacitors are referred to as *parasitic capacitors*. Although they have low capacitance values, (generally in the picofarad range) their presence can be troublesome in high-frequency circuits.

Power and telephone lines are also unintentional capacitors. There may be appreciable capacitance between two conductors or between one

Principles of Electrostatics

conductor and the earth. For short distances this capacitance may be negligible, but for long distances it becomes significant and must be considered when calculating line losses.

The human body is another unintentional capacitor. It forms one plate of a capacitor with the earth as the other plate. The body's capacitance depends on the amount and type of clothing and footwear the person is wearing, differences in floor materials, and the person's proximity to the ground. Capacitance measurements and estimates for the human body range from 80 to 500 pF.

Women have higher electrical capacitance than men because their shoe soles are usually thinner. However, tests have shown that 80 percent of all persons exhibit a capacitance value of 100 pF or less. While this value is small, it is a sufficient charge to cause extensive ESD damage to or destruction of sensitive electronic components. The DOD has chosen a capacitance value of 100 pF to represent the human body in its simulation circuit as discussed later in this chapter. However, other organizations have used alternative values, as discussed in chapter 9.

Capacitance Effects with Respect to Ground

The equation for capacitance can be rewritten as $Q = CV$. As before, Q is the charge, C the capacitance, and V the voltage. Because charge is directly proportional to the capacitance-voltage product, if charge is to remain constant while capacitance decreases, voltage must increase. Similarly, if the voltage decreases while the charge remains constant, the capacitance would have to increase. Therefore, if capacitance is decreased, voltage will increase until a discharge occurs in the form of an arc. For example, the charge potential of a common polyethylene bag may increase from a few hundred volts when it is lying on a bench to several thousand volts when a person picks it up because of the decrease in capacitance.

ELECTROSTATIC GENERATORS

Considerable time, effort, and expense have been put into the control and suppression of static electricity because of its potential for damage. Still, it is necessary to generate static electricity under controlled conditions for experimental and test purposes as well as practical applications. The most common systems for generating significant amounts of controlled static are the *Van de Graaff generator* and the *human body ESD simulator*.

Van de Graaff Generator

The Van de Graaff generator creates high-voltage electrostatic voltages of up to 6 million volts. The electric charge is stored at high potential

Electrostatic Generators

on a hollow metal sphere supported by an insulating column. The moving belt of silk or rubber is also housed within the column, as shown in FIG. 2-20. The belt, passing over an idler pulley at the top, is moved up the column by a motor-driven sheave. An electrode in the shape of a comb with pointed teeth is positioned near the bottom of the moving belt. It is connected by cable to a source of high voltage with respect to ground. The high-intensity electric field on the points produces inductive charge separation in the belt, removing electrons to the points by corona discharge and leaving the belt with a positive charge.

A second comb electrode, attached to the inside of the hollow sphere with a conductive rod, is positioned close to the top of the belt to collect the charge. Inductive charge separation in the metal sphere provides a high electric field strength near the points of the comb, and electrons pass to the belt by corona action, discharging the belt. The outside of the sphere is left positively charged because electrostatic charges always reside on the exterior of a conductor.

The charge energy gradually builds up on the sphere from the work done by the motor in driving the charged belt upward against the repulsive electric forces exerted by the similarly charged sphere. If the charge on the sphere reaches a saturation level with respect to the size of the sphere, it will repel additional charges and leak into the air.

A large Van de Graaff generator may have a 5-foot-diameter sphere mounted on a 50-foot insulating column. A machine of this size is likely to be equipped with a generator capable of 10-kilovolt (kV) output. These large machines are used to simulate lightning, operate high-voltage X-ray machines, and operate high-energy particle accelerators in nuclear fission experiments. Smaller benchtop units are used in classroom demonstrations of electrostatics.

In some manufacturing operations, fast-moving conveyor belts may inadvertently become Van de Graaff generators, creating high static charges on objects surrounding them. Care must be taken to ground or dissipate these charges to prevent electrical arcing, which could present a shock threat to persons nearby or trigger explosions and fires if flammable liquids or aerosols are nearby.

Human Body Simulation

People—not electrical machinery, power lines, or radio frequency sources—cause most of the damage and destruction to sensitive electronics by electrostatic discharge. ESD-sensitive electronic devices can even be destroyed by voltages that are below the average person's level of perception.

Equivalent circuits have been devised to simulate the charge, storage, and discharge characteristics of the human body. Many different

Principles of Electrostatics

electrostatic discharge simulators have been developed to determine damage and destruction thresholds for various vulnerable electronic components. They have also been employed to test the "ESD-proofing" of enclosed or packaged electronic products, instruments, and systems containing ESD-sensitive devices.

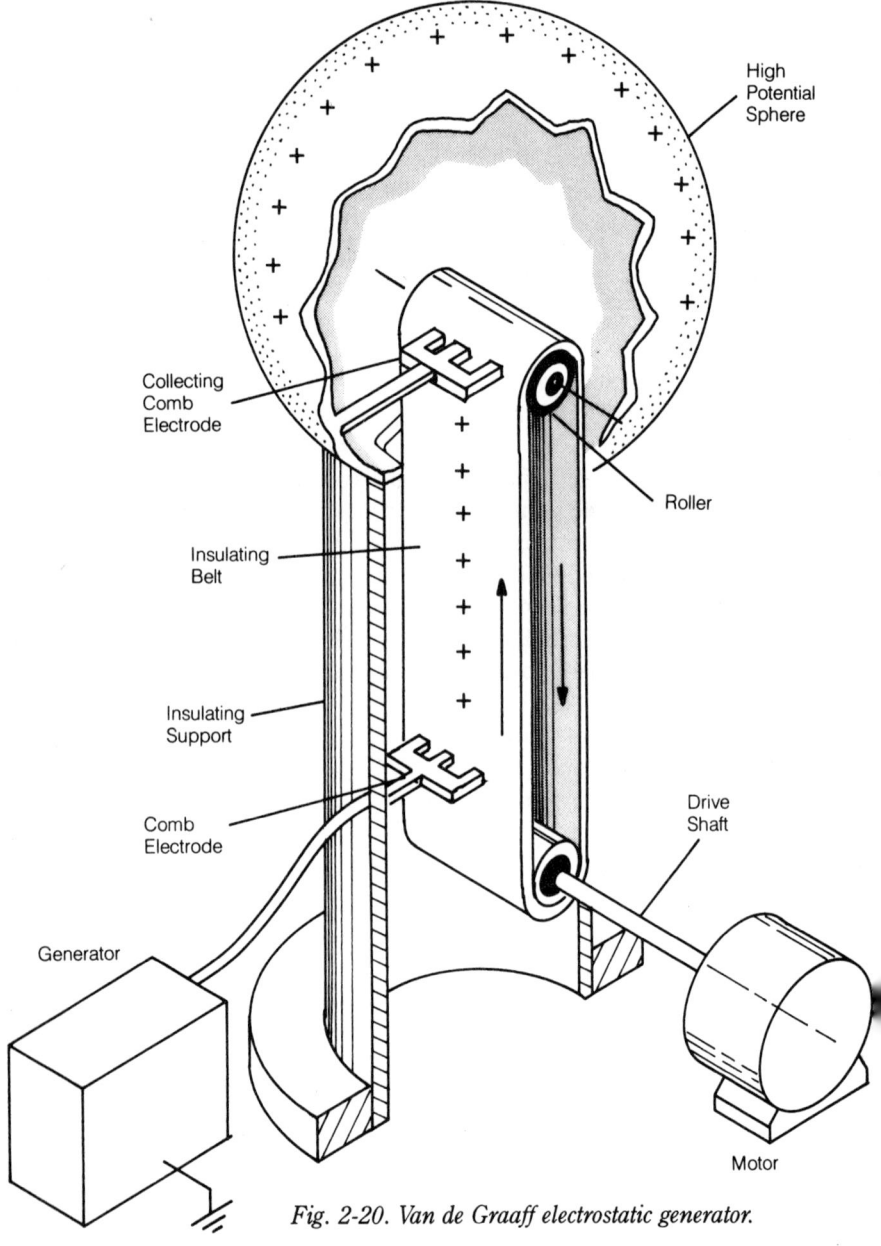

Fig. 2-20. Van de Graaff electrostatic generator.

Electrostatic Generators

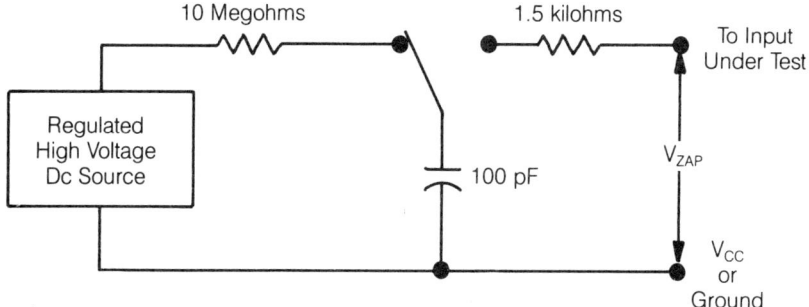

Fig. 2-21. Electrostatic discharge test circuit (human body model).

The most widely used human body simulators are based on a resistance-capacitance (RC) network that includes a resistor equivalent to human body resistance (50 to 5,000 ohms) and a capacitor equivalent to human body capacitance (80 to 500 pF). Figure 2-21 is the block diagram for the RC circuit specified in MIL-STD-883, Method 3015, to be used primarily for the testing and qualification of semiconductor devices.

The circuit has a current-limiting resistor to control the charging of the capacitor. The human body contains the sum of two capacitances: that due to the body's isolation from ground when the body is considered to be a sphere with the same surface area; that due to the parallel plate capacitance effect of the human body in contact with the ground plane through the soles of the shoes.

The switch shown for this simulator circuit is a bounceless, mercury-wetted relay or equivalent drawn in the charge position. The military standard diagram specifies a normally closed switch across the device under test (DUT) that is open during the discharge pulse and capacitance measurement, but this has been omitted to keep the diagram simple.

MIL-STD-883C calls out the current-limiting resistance R_1 to be between 10 megohms and 100 megohms and the body source resistance R_2 to be 1,500 ohms ±1 percent. Capacitor C_1 is specified as 100 pF ±10 percent based on the effective body capacitance. The high-voltage power supply must be capable of outputs from ±20 to 10,000 volts.

The waveform of the human body ESD simulator built in accordance with the military standard is shown in FIG. 2-22. This diagram defines the values for pulse rise time (T_{ri}) as less than 10 nanoseconds (ns) and pulse decay time (T_{di}) as 150 ±20 ns.

MIL-STD-883C defines the conditions for ESD step-stressing devices to a level of obvious failure. The 100 pF capacitor is charged gradually to increasing voltages, then discharged through the 1,500 ohms into the DUT, one pin at a time. Parametric tests, emphasizing leakage, are performed to check for out of tolerance conditions after each discharge. MIL-STD-883C is concerned with destructive, not latent, failures that may be generated at much lower levels.

Principles of Electrostatics

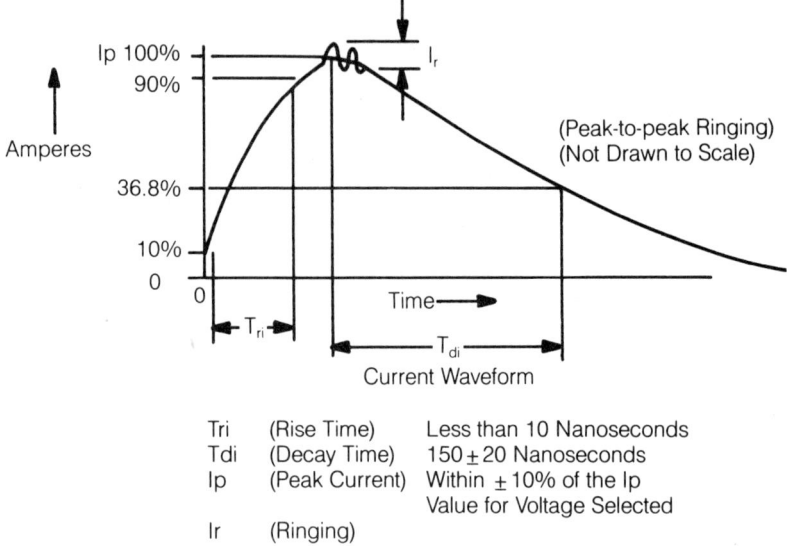

Fig. 2-22. Electrostatic discharge test circuit waveform (human body model).

Some ESD experts have been critical of the ability of simulators, based on the MIL-STD-883C circuit, to provide waveforms that are pure enough for meaningful, reproducible results. Their objections are based on studies suggesting that the rise time of human-body ESD could be less than 200 picoseconds, far shorter than the military standard.

Commercial human-body ESD simulators for testing semiconductor devices and other components are made as bench-type instruments with special fixtures for mounting the devices. However, hand-held instruments are better suited for testing computers, consumer products, or industrial equipment. Most simulators are based on the standard circuit or variations of it. They include variable power supplies and variable or replaceable resistors and capacitors so they can be organized in different circuit formats.

The military standard is binding only to companies that produce semiconductor devices for the military or are required by contract to use those devices. Other manufacturers may elect to use different simulators or component values that they believe are more appropriate to their test procedures. Chapter 9 in this book covers ESD simulation standards and simulators. It discusses alternate standards and describes typical simulation equipment available commercially.

Many in the industry believe that simulators should provide pulse rise times lower than the "less than 10 nanoseconds" of the military standard. Experiments have shown that ESD currents have rise times of 1 nanosecond or less. These rise times are said to be common at charge voltages less than 5,000 or 6,000 volts for relative humidities less than 35 to 40 percent. Nanosecond and subnanosecond rise times have been shown to be important in ESD-caused disruption of electronic equipment.

STATIC CHARGE DISSIPATION

As shown in FIG. 2-23, a static charge can be dissipated by surface or volume conduction to ground or by corona discharge into the surrounding air. The more effective these paths of conduction are, the faster the charge will dissipate. Corona discharge can be increased through ionization of the surrounding air and surface; volume conduction can be increased with antistatics that provide lower surface and volume resistivities.

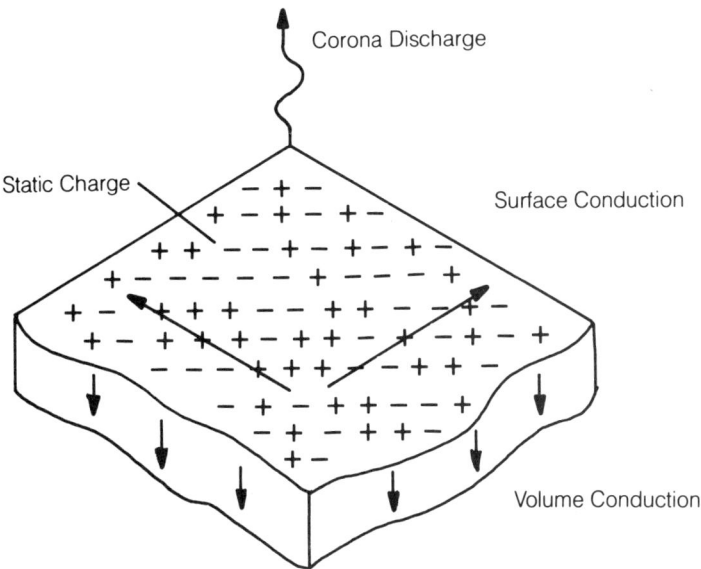

Fig. 2-23. Static charge dissipation methods.

Four basic techniques for solving ESD problems will be discussed in greater detail later in this book:

- Minimizing charge buildup
- Draining off charge
- Neutralizing charge
- Minimizing field and discharge effects

3

Electrostatic Discharge Damage Mechanisms

Certain types of electronic devices are susceptible to damage or degradation from ESD. The electrostatic charges referred to in this chapter are generated and stored on surfaces of ungrounded persons, on ordinary plastics, paper, glass, and other nonconductors, and on most common textile garments. ESD is the result of the discharge of this static charge. ESD also can be caused by other natural and man-made static generators.

Note that ESD control is a relatively new discipline. As a result, there is no universal agreement on standards, definitions, or even descriptions of failure mechanisms. As in many other areas of electronics, the military standards have assumed importance out of proportion to their mandated use because they act as unifying documents. They are not prepared unilaterally by U.S. government employees but are the result of a large cooperative effort on the part of concerned people both in government and industry.

In the absence of irrefutable scientific or engineering evidence that would discredit them, a DOD handbook and DOD and military standards have proved to be valuable references in ESD control. Many of the standards prepared by other agencies, both in the United States and other countries, are modifications or extensions of the U.S. military and DOD standards and handbooks. However, engineers and managers not legally bound to comply with the military and DOD standards have the option of using the sections that agree and substituting their own standards for those that do not agree.

SUSCEPTIBLE ELECTRONIC DEVICES

The packaged or unpackaged electronic devices subject to destruction by ESD include:

- All metal oxide semiconductor (MOS) devices, including NMOS, PMOS, and CMOS
- Junction field-effect transistors (JFETs)
- Bipolar digital and linear circuit ICs
- Operational amplifiers, monolithic microcircuits with MOS-compensating networks, on-chip MOS capacitors, or other MOS elements
- Gallium arsenide (GaAs) transistors and ICs
- Thin-film passive elements, including resistors, resistor networks, crystals, surface acoustic wave (SAW) filters, and signal conditioners
- Hybrid microcircuits containing any of the devices listed previously

Of the devices listed, most attention is being given to the latest-generation high-density CMOS ICs because of their economic importance and the rising demand for their use in electronics equipment. Loaded circuit boards, subassemblies, and even completely packaged instruments and systems containing ESD-sensitive devices are susceptible to damage because they are extensions of the individual devices they contain.

Well-designed, completely shielded equipment is less vulnerable than the individual devices or circuit boards containing those devices. Exposed conductors on the circuit boards can, in certain instances, act as antennas and make the devices even more susceptible to destruction or damage. However, damage at the enclosed product or system level is principally latent, typically showing up as leakage.

SEMICONDUCTOR DEVICE SUSCEPTIBILITY CLASSES

The destruction of or damage to a semiconductor device depends on its sensitivity or susceptibility to ESD. This sensitivity is the electrostatic voltage threshold; a voltage level that will cause catastrophic damage. TABLE 3-1 shows three classes of ESD sensitivity as defined by DOD-HDBK-263:

- Class 1: Devices susceptible to damage from ESD voltage levels of 1,000 volts or less
- Class 2: Devices sensitive to voltage levels of 1,000 to 4,000 volts
- Class 3: Devices sensitive to voltages of 4,000 to 15,000 volts

Electrostatic Discharge Damage Mechanisms

Table 3-1. Classification of ESD-Sensitive Parts (DOD-HDBK-263).

Class 1: Sensitivity Range 0 to ≤ 1000 Volts
Metal Oxide Semiconductor (MOS) devices including C, D, N, P, V and other MOS technology without protective circuitry, or protective circuitry having Class 1 sensitivity.
Surface Acoustic Wave (SAW) devices
Operational Amplifiers (OP AMP) with unprotected MOS capacitors
Junction Field Effect Transistors (JFETs) (Ref.: Similarity to MIL-STD-701: Junction field effect, transistors and junction field effect transistors, dual unitized)
Silicon Controlled Rectifiers (SCRs) with Io < 0.175 amperes at 100° Celsius (°C) ambient temperature (Ref.: Similarity to MIL-STD-701.
Thyristors (silicon controlled rectifiers)
Precision Voltage Regulator Microcircuits: Line or Load Voltage Regulation < 0.5 percent
Microwave and Ultra-High Frequency Semiconductors and Microcircuits: Frequency > 1 gigahertz
Thin Film Resistors (Type RN) with tolerance of ≤ 0.1 percent; power ≥ 0.05 watt
Thin Film Resistors (Type RN) with tolerance of > 0.1 percent; power ≤ 0.05 watt
Large Scale Integrated (LSI) Microcircuits including microprocessors and memories without protective circuitry, or protective circuitry having Class 1 sensitivity (Note: LSI devices usually have two to three layers of circuitry with metallization crossovers and small geometry active elements)
Hybrids utilizing Class 1 parts

Semiconductor Device Susceptibility Classes

Class 2: Sensitivity Range > 1000 to ≤ 4000 Volts

MOS devices or devices containing MOS constituents including C, D, N, P, V, or other MOS technology with protective circuitry having Class 2 sensitivity
Schottky diodes (Ref.: Similarity to MIL-STD-701: Silicon switching diodes (listed in order of increasing trr))
Precision Resistor Networks (Type RZ)
High Speed Emitter Coupled Logic (ECL) Microcircuits with propagation delay ≤ 1 nanosecond
Transistor-Transistor Logic (TTL) Microcircuits (Schottky, low power, high speed, and standard)
Operational Amplifiers (OP AMP) with MOS capacitors with protective circuitry having Class 2 sensitivity
LSI with input protection having Class 2 sensitivity
Hybrids utilizing Class 2 parts

Class 3: Sensitivity Range > 4000 to ≤ 15,000 Volts

Lower Power Chopper Resistors (Ref.: Similarity to MIL-STD-701: Silicon Low Power Chopper Transistors)
Resistor Chips
Small Signal Diodes with power ≤ 1 watt excluding Zeners (Ref.: Similarity to MIL-STD-701: Silicon Switching Diodes (listed in order of increasing trr))
General Purpose Silicon Rectifier Diodes and Fast Recovery Diodes (Ref.: Similarity to MIL-STD-701: Silicon Axial Lead Power Rectifiers, Silicon Power Diodes (listed in order of maximum, dc output current), Fast Recovery Diodes (listed in order of trr))
Low Power Silicon Transistors with power ≤ 5 watts at 25°C (Ref.: Similarity to MIL-STD-701: Silicon Switching Diodes (listed in order of increasing trr))
Thyristors (bi-directional triodes), Silicon PNP Low-Power Transistors (Pc ≤ 5 watts @ T_A = 25°C), Silicon RF Transistors)
All other Microcircuits not included in Class 1 or Class 2
Piezoelectric Crystals
Hybrids utilizing Class 3 parts

Electrostatic Discharge Damage Mechanisms

Note that the classification of a device depends on the presence of protective circuitry. That is, circuitry built into the device to protect it to the upper limit of a given classification upgrades the device to the next highest level. Protective circuitry is discussed in detail in chapter 4.

Not all standards and handbooks issued by the U.S. government agree. For example, MIL-STD-883C also has its own device ESD failure threshold classification:

- Class 1: 0 volt to 1,999 volts
- Class 2: 2,000 volts to 3,999 volts
- Class 3: 4,000 volts and above

Therefore, any discussion of ESD device failure threshold classification should specify the appropriate reference document.

TABLE 3-2 expresses ESD sensitivity according to device type, rather than class. All values listed are assumed to be voltage thresholds or limits for destruction. This table shows that some devices need more protection from ESD than others.

Table 3-2. Susceptibility of Various Electronic Devices Exposed to ESD. Courtesy 3M

Device Type	Range of ESD Susceptibility (Volts)
VMOS	30 — 1,800
MOSFET	100 — 200
GaAsFET	100 — 300
EPROM	100
JFET	140 — 7,000
SAW	150 — 500
Op-amp	190 — 2,500
CMOS	250 — 3,000
Schottky diodes	300 — 2,500
Film resistors (thick, thin)	300 — 3,000
Bipolar transistors	380 — 7,000
ECL	500* — 1,500
SCR	680 — 1,000
Schottky TTL	1,000 — 2,500

*PC board level

The failures resulting from ESD can be classified as: hard (catastrophic) or soft (upset); immediate or delayed; permanent or temporary; or direct or indirect. However, there is some overlap in these classifica-

Hard Semiconductor Failures

tions. The language used to describe ESD-related failures and malfunctions has not been standardized, so the same phenomenon may be called by different names in different references.

ESD pulses cause two general kinds of failures in semiconductor devices: hard and soft.

Hard failure is usually catastrophic, immediate, permanent, and direct. Most of the common hard or catastrophic ESD-related failures occur in unmounted diodes, transistors, and integrated circuits by three different mechanisms: thermal breakdown (or dielectric burnout), dielectric breakdown, and metallization melt.

A *hard failure* is an irreversible event requiring the replacement of the faulty component before the equipment can be operated again. These catastrophic failures are usually immediate, but they can also be latent. They can occur at any time, in or out of circuit boards, whether the equipment is operating or not. If the equipment is operating, operation can be normal or stressed: brownouts, surges, and other abnormal input conditions.

A soft failure, also called an *upset* is usually delayed, or *latent*, temporary, and indirect. It is a characteristic of integrated circuits and does not occur in discrete diodes, transistors, and other components.

In soft, or degradation, failure a variable of the component has shifted outside its specified range. Soft failure occurs only when the host system is operating. It may show up as a loss of information in a digital system, called *lockup*, or as a temporary malfunction in the operation of an analog circuit. Soft failures are not likely to cause hardware damage. In the case of digital computer circuits, proper operation is usually restored after lockup by reentering the data, resequencing the system, or powering down to clear it. By contrast, normal operation usually returns automatically in analog circuits.

HARD SEMICONDUCTOR FAILURES
Thermal Secondary Breakdown

In *thermal secondary breakdown*, also known as *junction burnout* or *avalanche degradation*, a hot spot is formed in a semiconductor pn junction by a constant voltage pulse, typically reverse biased. If a strong electrical transient from an ESD pulse of sufficient magnitude and duration is incident on a junction, it can cause secondary thermal breakdown that will melt a portion of the junction. These thermal breakdowns usually occur in constricted parts of the bipolar transistor. Some researchers see this form of destruction as a current-induced failure. Degradation or destruction is believed to be caused by discharge current over 2 to 5 amperes. Operational amplifier ICs are most vulnerable to current-induced failure.

Electrostatic Discharge Damage Mechanisms

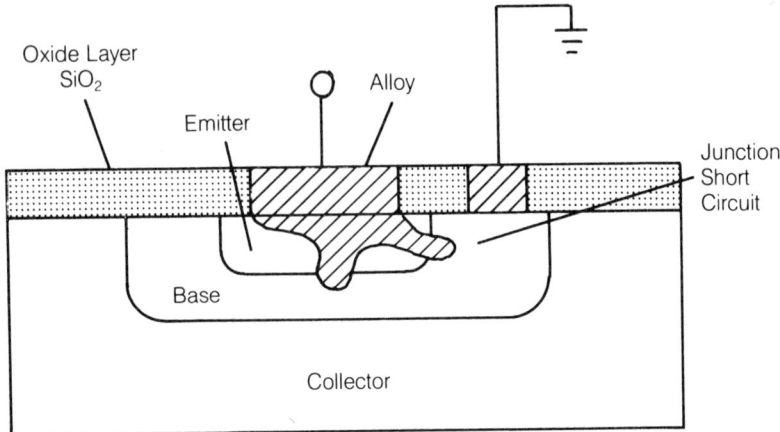

Fig. 3-1. Junction burnout in bipolar structure as a result of electrical overstress.

Figure 3-1 is a cross section of a bipolar transistor showing the emitter-base junction, which is most susceptible to damage. The emitter-base junction is the smallest of all junctions and is subjected to the highest power densities. This form of damage can occur in pn and pin diodes, as well as JFETs.

The temperature builds up at constrictions because the heat cannot dissipate in the time duration of the ESD pulse. As the temperature rises in the hot spot, the resistance falls and the current increases until the silicon in the hot spot melts. This melting short-circuits the junction and causes the device to fail.

Dielectric Breakdown

Dielectric breakdown, also known as *oxide punchthrough*, is the primary cause of failure in MOS semiconductors. Considered to be a voltage-induced failure, as opposed to a current-induced failure, it occurs when the voltage applied across the dielectric exceeds the breakdown voltage of the material. For MOS ICs, this dielectric typically has a breakdown voltage of approximately 8×10^6 volts per centimeter; this converts to 80 volts or less per 1,000-angstrom dielectric.

Figure 3-2 is a cross section of a MOS transistor showing the silicon dioxide dielectric layer and the site of a typical puncture. When the dielectric is punctured, current flows, destroying the oxide and forming a short circuit. The heat generated by the punchthrough might be sufficient to draw molten metal into the puncture cavity, creating a permanent and irreversible short. In some cases, however, the ESD pulse might not have

Hard Semiconductor Failures

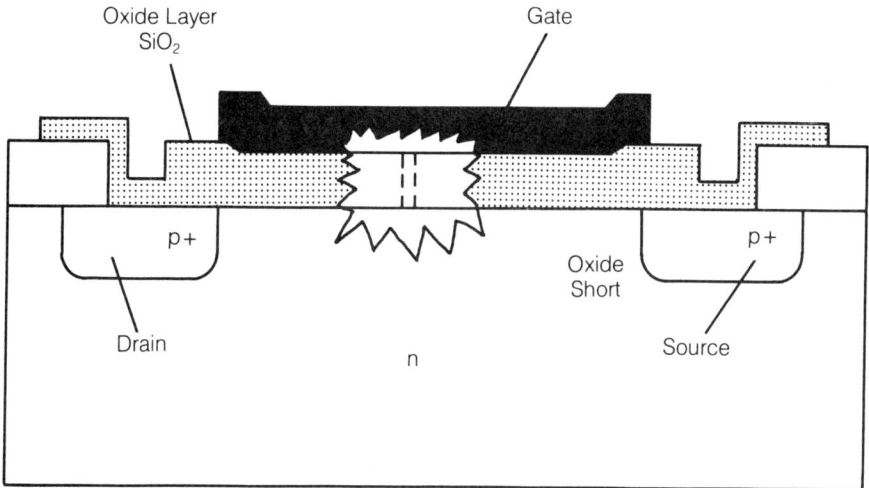

Fig. 3-2. Oxide punchthrough in MOS structure as a result of electrical overstress.

sufficient energy to short the oxide; instead the device might recover or heal from this puncture and continue to operate. Nevertheless, it remains predisposed toward later failure in normal operation.

This type of failure can usually be detected by measuring decreased breakdown voltage or increased leakage current. In addition to discrete MOSFETs and MOS and CMOS ICs, punchthrough can damage MOS capacitors, as well as both bipolar and MOS transistors with metallized crossovers.

Semiconductor manufacturers now add input protection devices and circuits to MOS ICs to make them less vulnerable to ESD (see chapter 4). Although limited protection is provided for the gate, it is often obtained at the cost of damage to or destruction of the protection network. Current-limiting resistors or protection diode junctions can fail due to thermal damage, and an input MOS transistor can be damaged by dielectric breakdown.

Metallization Melt

Metallization melt, or *burnout*, occurs when ESD transients raise the temperature of metallized paths or lines of semiconductor devices enough to melt the metal or fuse bond wires. Nonuniformity in conductors causes localized hot spots because of current crowding. Damage from this mechanism shows up as an open circuit in the device, as shown in FIG. 3-3. This damage can be caused by a semiconductor short circuit from a second breakdown or a gate oxide punchthrough which raises the current and melts the metal traces.

Electrostatic Discharge Damage Mechanisms

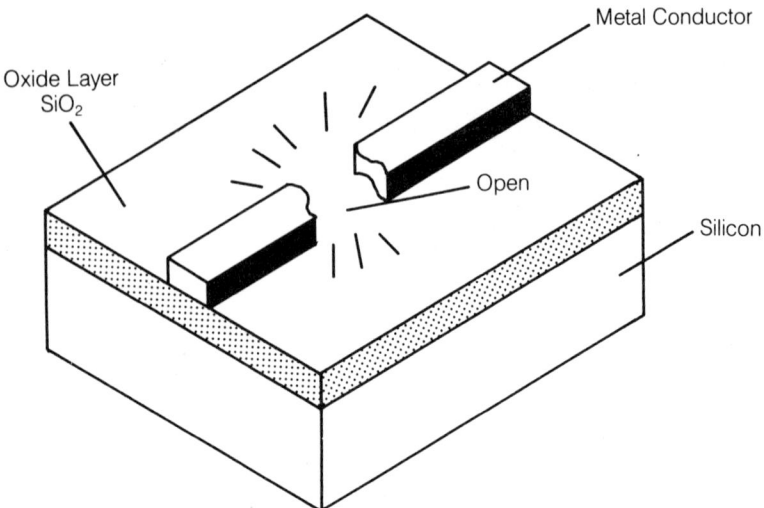

Fig. 3-3. Metallization melt as a result of electrical overstress.

ELECTRON MICROSCOPE ANALYSIS

There is a need for proof that one of these failures has occurred. One common technique is to observe the electrical characteristics (current-voltage) of a suspected microcircuit on a curve tracer. Figure 3-4 shows the characteristics for a shorted junction, a degraded junction, and a normal junction.

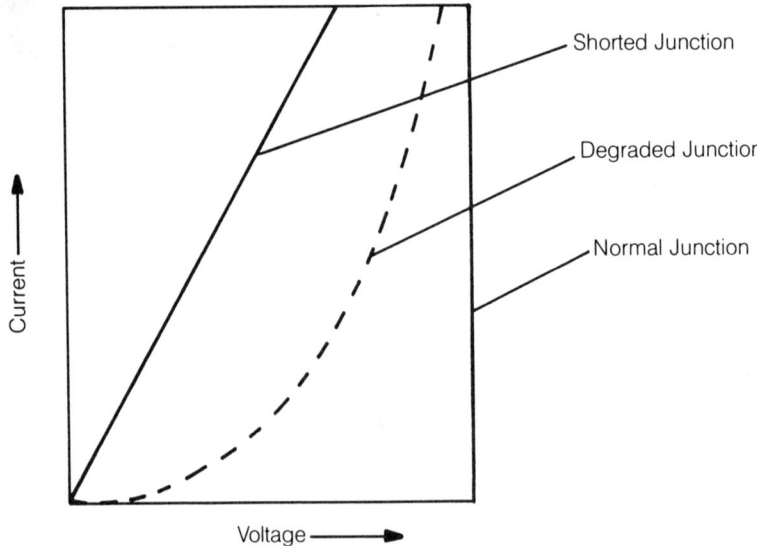

Fig. 3-4. Junction current-voltage trace.

Electron Microscope Analysis

Further proof can be obtained by examining the device with a scanning electron microscope (SEM). With this instrument, it is possible to see faults that are not visible with a conventional binocular microscope. The packaging or protective cases must be removed from the devices before they are examined by the SEM.

Figure 3-5 shows how ineffective packaging permitted ESD to destroy a 3N157 MOSFET. The image is enlarged 7,000 times. Figure 3-6 shows the destructive effect of ESD on an input pull-up resistor on a MOS character generator IC that experienced punchthrough. This image is enlarged 3,000 times. Figure 3-7 shows a detailed view of a 6-micron-diameter (0.0002-inch) hole punched in the aluminum metallization and silicon dioxide substrate of an operational amplifier IC. This view is enlarged 1,000 times and the second is enlarged 5,000 times.

Fig. 3-5. Destructive effect of ESD on a MOSFET magnified 7,000 times.

SOFT SEMICONDUCTOR FAILURES

A soft ESD failure or upset of an IC in host equipment can be caused by a spark discharge in the vicinity of the equipment. Electromagnetic

Electrostatic Discharge Damage Mechanisms

Fig. 3-6. Destructive effect of ESD on an input pull-up resistor on a PMOS integrated circuit magnified 3,000 times.

radiation generated by the arc and received by the equipment circuitry can create erroneous signals. One example is a lockup or equipment malfunction requiring a reset or power down to clear. Under certain circumstances, wait and disable states can occur, suspending equipment operation.

Capacitive or inductive coupling of an ESD pulse or direct discharge of an ESD pulse through a signal path also can cause an erroneous signal. Most discrete protective devices mounted on circuit boards in equipment are designed to block averaged signals. They may, for example, protect against lightning transients or spurious power line surges, but they might not be able to react fast enough to shunt the spikes of an ESD pulse.

Those semiconductors most susceptible to soft errors are ICs in semiconductor logic families that switch states with small amounts of energy or small voltage changes in high-impedance lines. These families include NMOS, PMOS, CMOS, low-power TTL, and GaAs. Linear circuits with high-impedance and high-gain inputs and radio-frequency (rf) amplifiers and other rf devices are also susceptible. Poor system design contributes to high ESD sensitivity.

Soft Semiconductor Failures

Fig. 3-7. Destructive effect of ESD on an operational amplifier as shown by a 6-micron (.0002-inch) diameter hole punched in aluminum metallization and silicon dioxide substrate magnified 1,000 times (A) and 5,000 times (B).

4
Semiconductor Device Board and System Protection

There are three ways to prevent damage and destruction of electronics products by ESD: prevent ESD completely; increase the inherent ESD immunity of devices and circuits; and prevent the coupling of ESD into devices and circuits. Since inherent ESD immunity of devices and circuits is a function of their design, this chapter considers only the principles involved in increasing ESD immunity. No design details are provided.

CONDITIONING THE ENVIRONMENT

Many important steps can be taken to reduce the threat of ESD in the factory or other controlled environment where sensitive devices and circuit boards are stored, tested, inspected, packaged, and assembled. Such steps range from maintaining minimum levels of relative humidity in the room to using ionizers, conductive floors, conductive benchtops, conductive containers, and personnel ground straps. These methods are all discussed in chapters 5 through 7 of this book. The objective of all these provisions is:

- Minimizing charge build up
- Draining charge
- Neutralizing charges
- Minimizing field and discharge effects

PROTECTION FOR INTEGRATED CIRCUITS

The inherent immunity of a device to destructive damage from ESD can be improved with protection circuits. For more than ten years, semiconductor manufacturers have been building ESD-protective networks into their ICs, particularly MOS devices. But these devices or networks have not been sufficient for complete protection, and they are normally supplemented with protective networks on the host circuit board. The chip-protection circuits, enabled when ESD voltage exceeds a preset limit, provide alternate paths for the safe flow of static charges to ground.

With carefully designed networks of resistors, diodes, and capacitors, manufacturers have been able to raise the withstand voltages of some devices from less than ± 100 volts to more than $\pm 2,000$ volts. However, even this level is well below the 4,000- to 15,000-volt range that can be built up when a person walks across a room—depending on humidity and surface conditions.

The most vulnerable devices are the insulated gate structures in MOS ICs that are subject to punchthrough. CMOS ICs are likely to be protected only to about 1,000 volts, and many have less protection.

One method of punchthrough protection is the use of solid-state, current-limiting, and voltage-clamping devices on the IC chip between critical points, such as input and ground. If the ESD-induced voltage exceeds the clamping level, the clamping element turns on and becomes a low-impedance shunt. Several current-clamping devices can be cascaded so that the faster, lower-rated device turns on first and handles high currents until the second, more rugged, device turns on.

Regardless of arrangement, the impedance of the protection shunt path must be low enough so that the voltage developed across this impedance will not exceed the breakdown voltage of the transistors on the IC. As IC geometries becomes smaller and less able to withstand ESD currents and voltages, the design of on-chip IC protection circuits becomes more complex and critical.

CMOS devices have been used for many years in applications in which the primary concerns were low power consumption, wide power supply range, and high noise immunity. However, the metal-gate CMOS is too slow for many applications. Where high-speed integrated circuits were needed for applications such as microprocessor memory decoding, it was necessary to use the faster bipolar families such as the Schottky LS TTL, which meant sacrificing the best qualities of CMOS.

To overcome this handicap, a new family of CMOS devices was introduced for high-speed applications, while retaining all the advantages of CMOS. The evolutionary process of converting the CMOS is illustrated in

Semiconductor Device Board and System Protection

FIG. 4-1. The high-speed CMOS device shown in cross section as FIG. 4-1B is about half the size of its metal-gate predecessor (FIG. 4-1A), thus saving significant silicon chip area.

The silicon gate process allows smaller gate or channel lengths because the gate is used in a self-aligning process. The gate defines the channel during processing, eliminating registration errors and the need for gate overlaps that can be seen in the metal-gate CMOS cross section. Gate capacitance is lowered significantly by the elimination of gate overlap, giving the IC higher speed. The smaller gate length also results in higher drive capability per unit gate width, ensuring more efficient use of chip area.

All MOS devices have insulated gates that are subject to voltage breakdown. The gate oxide for metal-gate CMOS devices is about 800 angstroms thick and breaks down at a gate-source potential of about 100 volts. To guard against this breakdown from ESD or other voltage transients, the protection network shown in FIG. 4-2 is used on each input to the CMOS device.

The high-speed silicon-gate CMOS, like all MOS devices, also has an insulated gate that is subject to voltage breakdown. The gate oxide for these devices also breaks down at a gate-source potential of about 100 volts. Device inputs for this faster CMOS with higher gate density are protected by resistor diode networks, such as the one shown in FIG. 4-3. With this circuit, the device inputs typically can withstand a discharge greater than 2,000 volts when tested with the simulator circuit illustrated in FIG. 2-21. Although these input protection networks provide significant protection, no CMOS devices are immune to the large ESD discharges that can be generated during handling. As a result, all reasonable precautions must be observed.

In general, most IC protection networks are built into the input leads, with no protection at the output leads. Yet, IC destruction and degradation from ESD entering the output leads has been reported. To correct this condition, some manufacturers have included both input and output protection networks for their ICs.

One serious limitation of on-chip protective networks is that failures in these networks can disable the rest of the device and make its replacement necessary. The diodes, resistors, and transistors built into the protective network do not participate in the functioning of the chip, but if they are destroyed by an overvoltage while the rest of the circuit remains intact, the complete device still must be replaced.

In the ideal on-chip ESD-protection circuit, high voltages see a low-impedance path around a gate, but low voltages are not affected by this path. The circuit has a fast response between high- and low-impedance conditions, yet it does not degrade the speed of the device it protects. It provides the clamping shown in FIG. 4-4.

Protection for Integrated Circuits

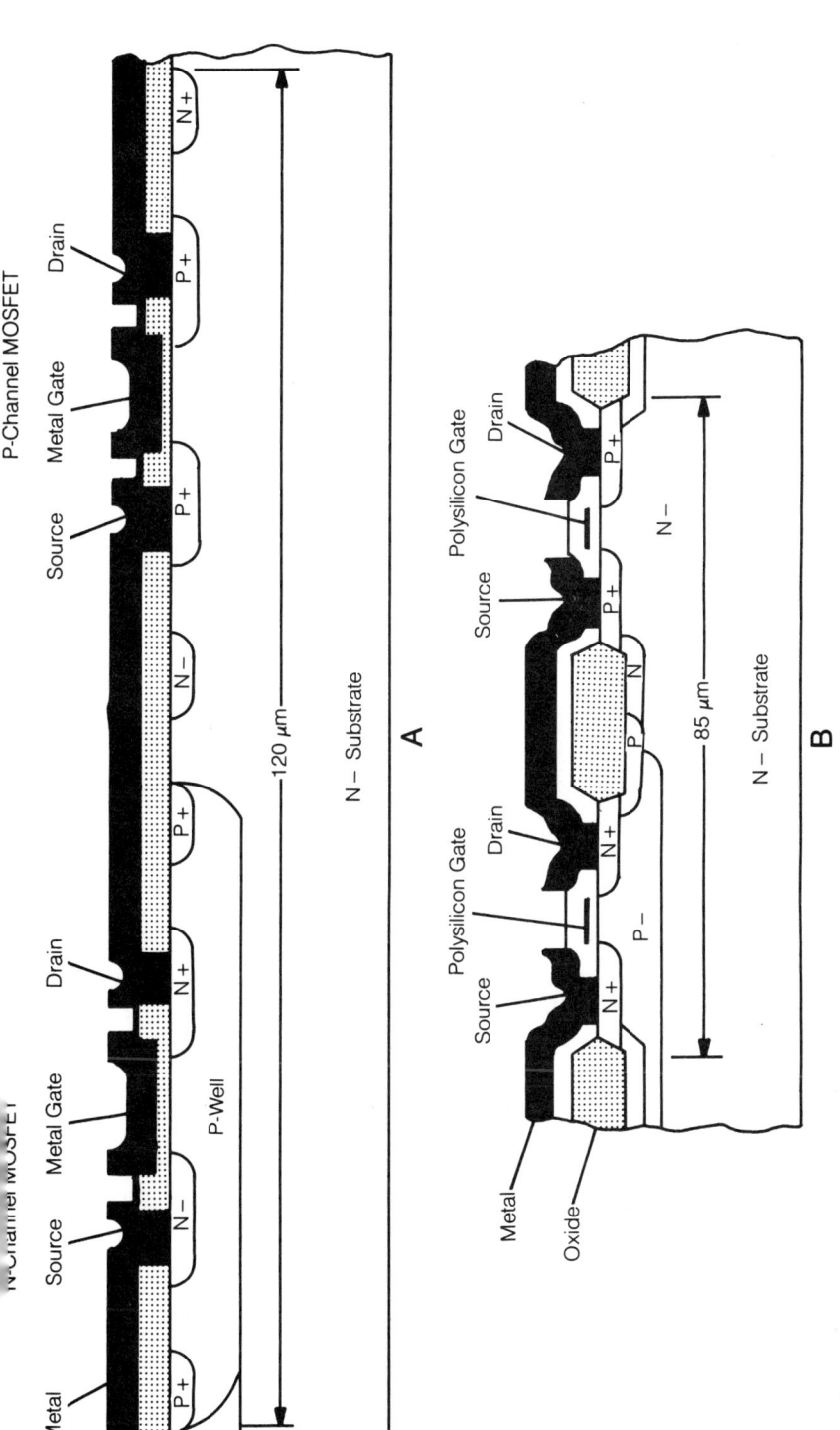

Fig. 4-1. *The evolutionary process in reducing the cross section of metal-gate CMOS (A) to high-speed silicon gate CMOS (B).*

Semiconductor Device Board and System Protection

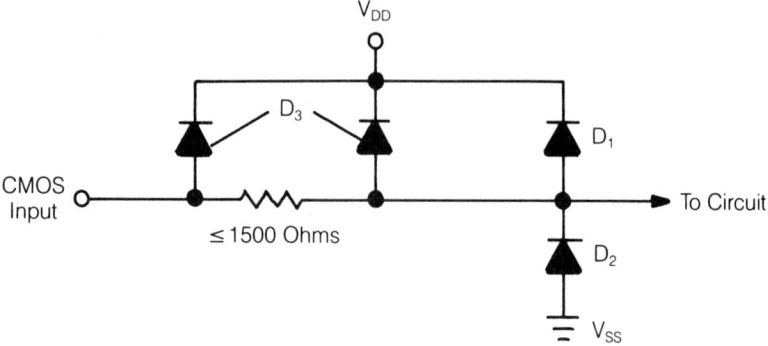

Fig. 4-2. Input protection network for metal-gate CMOS.

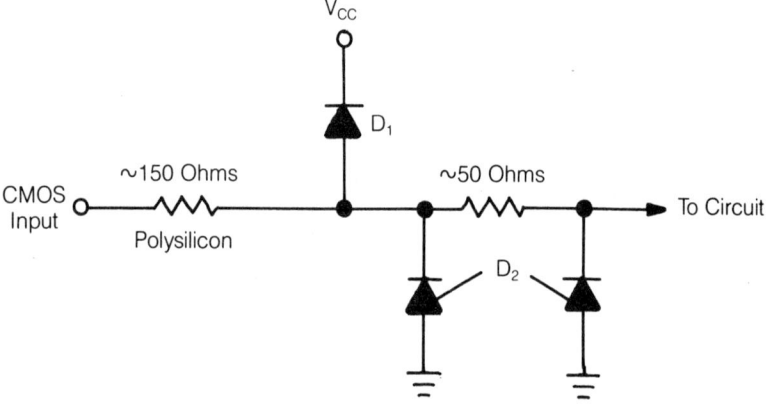

Fig. 4-3. Input protection network for high-speed silicon-gate CMOS.

Unfortunately, on-chip protective networks can adversely affect the performance of the devices they protect. They also use much space on the silicon substrate, and provide little protection (i.e., only up to 1,000 volts).

PROTECTION FOR SEMICONDUCTORS AT BOARD LEVEL

For many years, circuit-board manufacturers have been successful in protecting ESD-vulnerable ICs with transient suppressors and protection networks mounted on the circuit board as well as integrating protection networks into the circuit chip. ESD energies are small. For example, 150 picofarads, charged to 25 kilovolts, stores less than 50 millijoules. Therefore, board-level protection must be fast, but it does not have to handle much energy.

Protection for Semiconductors at Board Level

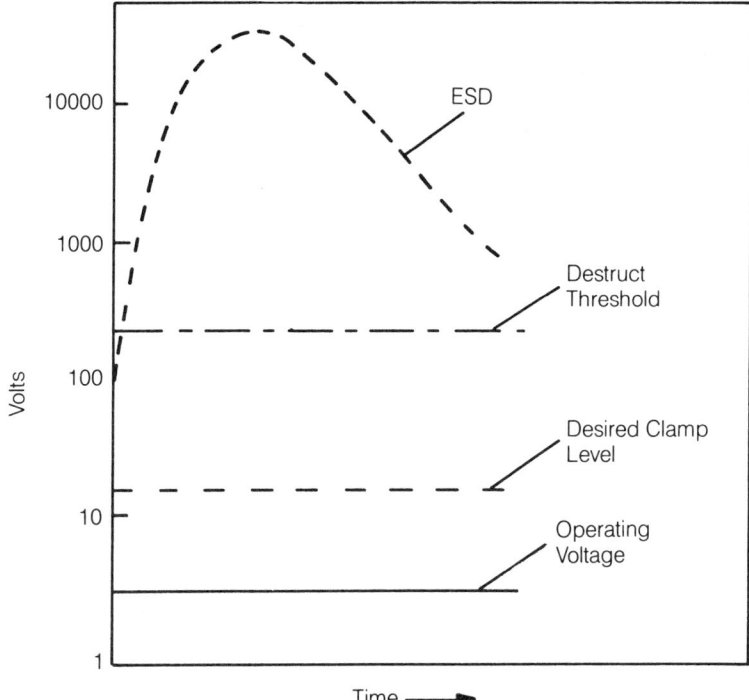

Fig. 4-4. Concept of clamping for ESD protection.

Figure 4-5 shows two possible networks using a series resistor to reduce ESD damage. These networks are useful in protecting digital inputs and outputs, analog inputs and outputs, three-state outputs, and bidirectional input/output (I/O) ports. Figure 4-5A shows the advantage of requiring minimal board space. Its disadvantage is that it can severely affect rise and fall times, propagation delays, and output drives. By contrast, FIG. 4-5B calls for more circuit-board area and costs more, but the impact on ac and dc characteristics is minimized.

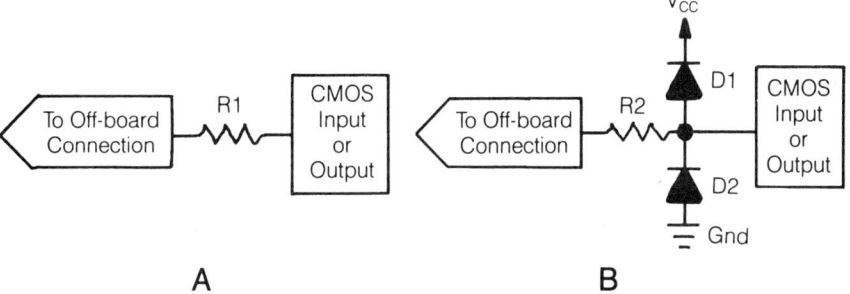

Fig. 4-5. Networks for minimizing ESD on circuit boards.

Semiconductor Device Board and System Protection

Transient suppressors also can be mounted on the host circuit board to limit the passage of transients before they reach a vulnerable integrated circuit. Two common types of suppressors are: silicon transient voltage suppressers (TVSs), and metal-oxide varistors (MOVs). Both types are packaged for conventional leaded- or surface-mounting on circuit boards. The current through these devices is extremely small at low voltages. At voltages above their specified clamping ratings, the current increases rapidly, the voltage drops, and the source impedance cuts off or clamps the voltage spike.

The TVS is a zener-type pn junction diode designed for overvoltage protection. Because of its reverse-bias voltage-clamping characteristic, the diode will break down and become a short circuit when the applied voltage exceeds its rated avalanche level. When this applied voltage (reverse bias) falls below the zener breakdown level, current is restored to its normal saturation level.

Response times for TVSs are measurable in picoseconds, and they are primarily intended for use in dc circuits. However, two TVSs can be placed back-to-back to protect ac lines. Some manufacturers offer this configuration in a single package.

The three most important considerations in selecting a TVS are the ratings for pulse power (peak pulse current multiplied by the clamping voltage), standoff voltage, and maximum clamping voltage. TVSs clamp more sharply than MOVs and also have a capacitance of about 1,000 to 2,000 pF at zero bias. These parasitic capacitances might not be acceptable in some systems.

The MOV is a variable resistor intended primarily to protect against voltage transients in the ac line. It behaves like two back-to-back TVSs. As nonlinear resistors with resistance changes that are a function of the applied voltage, MOVs have bilateral and symmetrical characteristic curves. As a result, the circuit is clamped in both the positive and negative directions.

When an input voltage transient greater than 4 volts rms appears across the MOV, its resistance drops sharply, so it effectively becomes a short circuit to bypass the transient. The device also has the ability to absorb more current flowing through the device at the time of the transient than does a TVS.

MOVs are made of powdered metal oxide, principally zinc oxide, mixed with binders and pressed into disks, blocks, or cylinders. When fired at temperatures above 1,000°C, the pressed slugs become a matrix of conductive metallic oxide grains separated by highly resistive boundaries.

The electrical properties of the MOV are determined by the volume of its slug; it absorbs energy uniformly throughout its volume, rather than only at the junction as does the TVS. MOVs have high short-term energy absorption capability, but the amount of energy absorbed is limited by slug volume. Although power absorption is not of prime concern in ESD protection, the MOV can handle transients from other man-made and natural sources as well. There are differences in the characteristics of MOVs and TVSs, but they can be combined in the same circuit for added protection without mutual interference.

MOVs function as capacitors with values of 5,000 to 10,000 pF. However, these parasitic capacitances, far larger than those of the TVS, also might make them unacceptable in some systems. Care must be taken to ensure that the MOV's high capacitance will not interfere with normal circuit operation.

In designing circuit boards, care must be taken to prevent radiated ESD from coupling to the electronic circuits. The conductive paths on circuit boards can act as receiving antennas for ESD-generated fields. To minimize coupling to these antennas, line lengths are kept short and loop areas are kept small. Lines longer than 1 inch and loop areas larger than a fraction of a square inch can receive significant amounts of radiation.

The primary means for preventing ESD upset is the use of filters designed into circuit boards to prevent ESD from being coupled into devices. Shunt-capacitive filters, series-inductive filters, or combinations of the two can be used. Filter elements must be located within 1 or 2 inches of the device pin or radiated ESD might bypass the filter.

SYSTEM IMMUNITY TO ESD

A technical discussion of engineering design techniques for ESD system immunity is beyond the scope of this book; however, some general guidelines apply to circuit board filters. For example, ESD filter capacitors should have values less than a few hundred picofarads to keep their stray inductance low. This ensures low capacitor impedance at ESD frequencies that can exceed 1 GHz. In addition, inductive elements are typically ferrite beads because wirewound inductors have too much stray capacitance. Ferrite filters must filter effectively at frequencies of several hundred megahertz or higher.

If the input has a low impedance, a series ferrite inductor will be an effective filter because it will form a voltage divider with the input impedance. On the other hand, if the input impedance is high, a shunt-capacitive filter will be effective because its low impedance will bypass the high input impedance.

Although they are not filters, the discrete high-speed clamping transient suppressors mentioned earlier can be used in place of shunt capacitors because they can switch in nanoseconds. There might be enough stray capacitance and inductance in some circuit boards to act as the only filter needed.

Semiconductor Device Board and System Protection

The use of enclosures is based on their ability to prevent both radiated and conductive coupling of ESD noise to circuits. A person charged to 20 kV can produce a spark up to ³/₄ inch long. Either insulating or conductive barriers can prevent these arcs from reaching susceptible electronic devices. Both will prevent conductive coupling, but an insulating layer will not prevent radiated ESD from coupling to the devices.

Complete metallic enclosures will shield the circuitry from most radiated effects of ESD, but secondary arcs from the shield to the circuit can be coupled to the devices. These arcs can be prevented by separating the shield from the circuit and, if possible, grounding the circuit. Some enclosures are made of insulating materials but have conductive metal shields inside.

Most enclosures are not complete shields because they have holes, air outlets, or conductive hardware penetrating the barriers. These paths allow ESD to pass through or around the shield. In properly designed enclosures, the holes are located away from susceptible circuits or devices, and they are made as small as possible; multiple holes, properly spaced, are preferable to one large hole.

Cables must be considered in achieving overall system ESD immunity. As the largest receiving antennas within most systems, cables may have large voltages and currents from radiated ESD noise. Radiated coupling is prevented by keeping cable and condutor lengths as short as possible and by minimizing loop areas. The circuit lines of the cable should be enclosed with conductive metal shielding terminated at both ends of the cable. Each active conductor in the cable should be as close as possible to its return line. Filtering concepts used in circuit design also apply in cable design. Some cable connectors have built-in shunt capacitors or clamping suppressors; others include series ferrite inductive beads.

Software solutions have been applied to ESD. It has been possible to write program subroutines that periodically refresh, check, and, if necessary, recover data. Refreshing is done when the data must meet specific conditions for the next operation to occur. Reconstruction is the main function of a recovery program. These subroutines can be programmed into read-only memory (ROM), giving the designer a firmware option.

5
Protective Materials and Packaging

All materials that contact or are close to ESD-sensitive products must be carefully selected for properties that minimize static charge buildup or drain off charge. This care applies to containers for holding parts temporarily during in-plant processing, such as tote trays and boxes, as well as containers used for long-term storage or shipment, such as protective bags, magazines and individual device shipping containers. Protective boxes can be lined with conductive foam; circuit boards containing sensitive devices or components can be protected with conductive shunts (shorting plugs) and cushioned with suitable conductive packaging material.

A wide selection of materials is available for the manufacture of packaging, and each material has its own set of physical properties. The selection of materials and packaging is conditioned by need and cost. Ideally, the cost of protective surfaces and packages should be in proportion to the level of protection required.

This chapter discusses the basic principles involved in the selection of packaging or materials that contact ESD-sensitive products. Major purchases of ESD-protective materials and packaging should not be made until the customer has thoroughly surveyed the materials and products available and is satisfied that the manufacturers' or suppliers' claims are valid. To be satisfied, the user might need to initiate an in-house product qualification program and periodic testing of lot samples.

ESD-protective containers must provide protection against: triboelectric generation, electrostatic fields, and direct discharge from contact with static-charged persons or objects. No single type of material satisfies

Protective Materials and Packaging

all the requirements for comprehensive protection, so it might be necessary to compromise in order to balance level of protection, cost, and physical properties such as flexibility and durability. In practice, some protective products, such as bags and work surfaces, contain combinations of different protective materials to achieve the desired protection and physical properties.

MATERIALS CLASSIFICATION

Static protective materials are classified in accordance with their range of surface resistivity, an inverse measure of the material's surface conductivity and volume resistivity an inverse measure of volume conductivity. The Department of Defense identifies three categories: conductive, static dissipative, and antistatic (DOD-HDBK-263). According to this classification, conductive materials have a surface resistivity of 10^5 ohms/square or less; static-dissipative materials have a surface resistivity between 10^5 and 10^9 ohms/square; and antistatic materials have a surface resistivity between 10^9 and 10^{14} ohms/square.

As stated earlier, there is no universal agreement on definitions in the field of ESD control. The Electronics Industries Association (EIA) does not agree with the DOD classification scheme, and it offers its own. EIA Standard RS-541 defines only conductive and dissipative materials, either in terms of surface resistivity or volume resistivity. This standard defines conductive materials as having either a surface resistivity of less than 10^5 ohms/square or a volume resistivity of less than 10^4 ohm-cm. Furthermore, it states that conductive materials are not necessarily antistatic. EIA defines static-dissipative material as having either surface resistivity from 10^5 ohms/square to 10^{12} ohms/square or volume resistivity from 10^4 ohm-cm to 10^{11} ohm-cm.

The EIA does not assign values of surface resistivity or volume resistivity to its definition of *antistatic*. That term applies to "those materials exhibiting antistatic properties which minimize the charge when rubbed against or separated from themselves or other similar materials." As a result, the degree of charge generation depends on the specific materials involved.

According to EIA Standard RS-541, materials can be naturally antistatic (intrinsic state) or can be made antistatic by one of two methods: compounding with additives to minimize charge generation (bulk treatment), or treating by spraying, dipping, printing, or wiping with a topical antistat (surface treatment).

The International Electrotechnical Commission (IEC) in its revision of IEC Publication 801-2 defines antistatic material as an ESD-protective material having a surface resistivity greater than 10^5, but not greater than 10^{11} ohms/square.

Materials Classifications

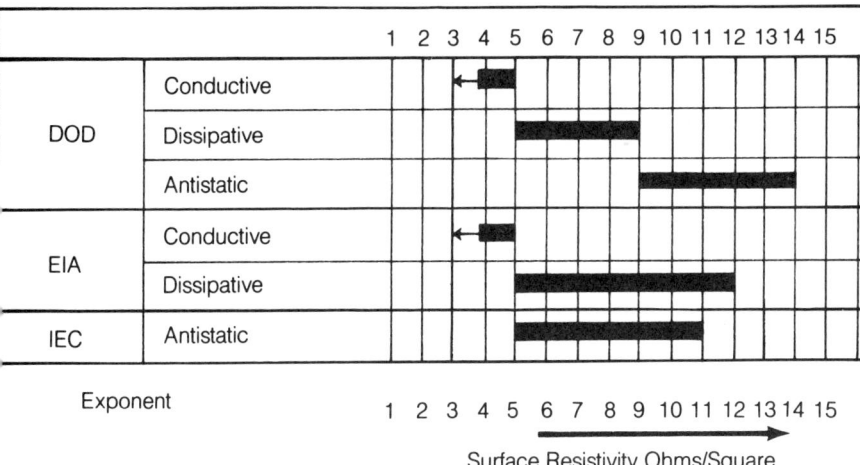

Fig. 5-1. Material classified by surface resistivity according to DOD, EIA, and IEC.

All types of conductive materials have proved to be useful in ESD protection, and it is important that the conductivity be appropriate for the level of protection desired. As a result, cost may be the key factor in selecting materials. In general, the cost of protective materials is proportional to conductivity.

Figure 5-1 is a plot of the limits of materials as classified in DOD-HDBK-263, EIA RS-541, and IEC 801-2. Despite this difference in definitions, it is important that all ESD-protective products are labeled with the range of values of conductivity, or that this information is available to the user. The user then can compare this data with any of the standards or definitions.

TABLE 5-1 compares the different properties of materials as classified by DOD-HDBK-263 qualitatively. There are no lines of distinction for applications purposes between any two adjacent classes of material in the table. For example, the properties of a conductive material at the high end of its resistivity range could be equivalent to the properties of a static dissipative material at the lower end of its resistivity range. Similarly, the properties of high-resistivity static-dissipative materials are comparable to those of low-resistivity antistatic materials.

Because triboelectric generation is a friction process, one of the principal ways to reduce it is to increase the material's *lubricity*, a measure of surface smoothness and the lubricating action of moistness. The higher the lubricity of the surfaces being rubbed, the lower the friction, and therefore, the lower the generated charges.

Antistatic materials are impregnated with *antistats*, chemicals that constantly migrate to the surface, forming a sweat layer that increases the

Protective Materials and Packaging

Table 5-1. Comparison of ESD-Protective Materials (DOD-HDBK-263).

Conductive	Static Dissipative	Antistatic
1. Could present a personnel safety hazard when contacting high voltages and hard grounds. 2. Could damage electrical circuitry of parts or assemblies during testing if electrical connections contact conductive surfaces. 3. Steels (except corrosion resistant) are prone to corrosion. Protective coatings such as paint will destroy the surface conductive properties and could be static generative. 4. Aluminum will form aluminum oxide on its surface reducing conductivity and increasing its ability to generate static. 5. Hard surfaces such as metal provide little protection from physical shock to items dropped thereon. 6. Materials should be reviewed for flammability, corrosivity, toxicity, bacterial growth, crumbling, powdering, shedding, flaking, brittleness, outgassing, long term chemical reaction with parts. 7. Protection against triboelectric generation depends upon the material used.	1. Presents the same hazards as listed under "Conductive" 1 and 2 except to a lesser degree. Hazards depend upon the magnitude of the voltages and the types of parts and circuits tested. 2. See item (6) under "Conductive." 3. See item (7) under "Conductive."	1. The effectiveness of hygroscopic antistatic materials are reduced in low relative humidities since their antistatic properties are dependent upon absorbing moisture from the air. 2. The accumulation of dirt, oils and silicone have an adverse effect on the antistatic properties of hygroscopic antistats. Cleaning with solvents such as alcohols, ketones and other hydrocarbon based solvents can remove the antistats. May require periodic treatment with a topical antistat. 3. Antistats used in some hygroscopic antistatic materials can track onto items and act as a foreign substance which could react with other materials adversely. This has been shown to be a problem with the lubricant in miniature bearings. 4. See item (6) under "Conductive." 5. Hygroscopic antistatic materials generally provide protection against triboelectric generation. The triboelectric generation characteristics of other antistatic materials depends upon the material used.

Materials Classifications

material's lubricity. These hygroscopic chemicals attract moisture from the surrounding air. If the relative humidity is low, their effectiveness will decrease. Periodic tests might be necessary to ensure that this property has not deteriorated. Most antistatic plastics have a tendency to lose effectiveness after prolonged contact with paper products or exposure to the air.

Conductive materials are typically impregnated or loaded with carbon to produce volume-conductive substances. These materials are widely used as electrostatic shields (Faraday cages) to minimize induced charge effects of electrostatic fields. Because the shielding ability of a material is directly related to its conductivity, antistatic materials (DOD definition) cannot be used as electrostatic shields.

Electrostatic shields also provide protection for ESD-sensitive electronic components against destructive electromagnetic pulses. These pulses might otherwise be induced in an enclosure from an ESD high-voltage spark. To be effective, an electrostatic shield must form a complete conductive cage around the component to be protected. An uncovered tote box does not form a complete cage and, therefore, cannot completely shield against an electrostatic field.

In addition to providing protection against electrostatic fields, many conductive and static-dissipative materials also provide protection from triboelectric generation. However, some metals have the ability to create significant static charges from triboelectric generation (see TABLE 2-3). Aluminum, for example, can generate substantial electrostatic charges when rubbed with a common plastic. Conductive materials distribute charges over their surfaces, but other materials rubbed against them (or from which they are separated) can become highly charged, particularly if they are insulators.

Whether conductive, static dissipative, or antistatic, ESD-protective materials can be formed, cast, stamped, or molded into various shapes. Metals, a prime example of conductive materials, can be cast, stamped, drawn, spun, or assembled and welded into most any shape. However, conductive, static-dissipative, and antistatic plastics can be molded or formed into sheets, tubes, or vessels. Fiberboard, melamine laminates, and similar materials in laminated or homogenous form can be assembled or folded and fastened into boxes and other useful container shapes.

The forms or shapes of ESD-protective materials include:

- Sheets and plates in various sizes and thicknesses for use as bench and table surfaces, flooring, floor mats, and covers
- Containers formed as trays, vials, carriers, boxes, or bottles.
- Rigid shorting bars and clips to short leads and terminals of ESD-sensitive circuit boards

Protective Materials and Packaging

- Foam sheets used to short sensitive component leads, and as assembly connectors and package cushioning
- Bubble-pack material or open-cell plastic, used to cushion and package, conforming to MIL-P-81997 and PPP-C-1842
- Flexible sheet materials used to fabricate pouches or bags, in accordance with MIL-B-82467
- Flexible fabric materials, used to manufacture seat covers and apparel (including smocks and gauntlets)
- Flexible stretch materials, used to manufacture gloves and finger cots
- Flexible bulk material, used to fabricate personnel shoe grounding straps
- Carpets and floor tiles with conductive, static-dissipative, and antistatic properties

The typical properties of conductive materials include:

- Opacity (except in the form of metallized plastic)
- Higher cost than antistatic materials
- Maintenance-free
- Constant surface resistivity for life of product
- Surface resistivity independent of moisture or relative humidity
- Ability to accept direct label printing

The typical properties of antistatic materials include:

- Opacity or translucence
- Lower cost than conductive materials
- Necessity of periodic tests to determine functionality
- Dependence on relative humidity to perform
- Inability to accept printing

There might be significant quality variations in antistatic packaging materials. Since static decay times of materials vary among lots at an individual manufacturer, they might vary dramatically among different manufacturers.

All protective packaging must have an obvious label stating that ESD-sensitive components may be inside. ESD-sensitive components must not be removed from their containers except in static-free workstations, and then only when work is to be done on them. Before removing a component from its protective container, the operator should: place the container on a conductive grounded bench mat or surface, be sure the wrist strap fits snugly and is properly plugged into the ground receptacle, and touch hands to the benchtop.

PROTECTIVE BAGS

Laminated plastic bags containing flexible metallic film or foil shields are alternatives to rigid, metal containers. The metal in the walls of the bag acts as a Faraday cage (see FIG. 5-2). These bags are easier to handle than metal boxes, cost less, and will not damage their contents. These bags, either opaque or transparent, are used for shipping and storing components and circuit boards.

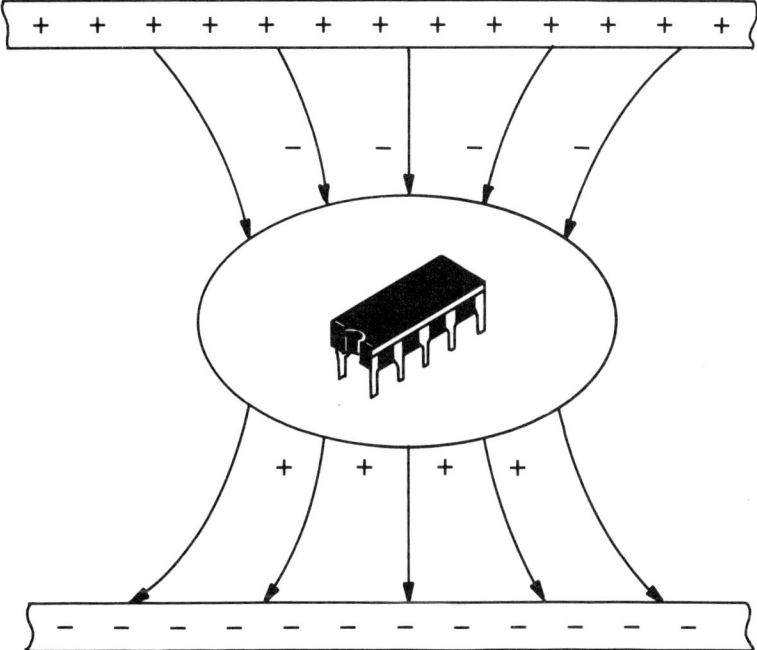

Fig. 5-2. No electric field exists inside a Faraday cage.

The metallized barrier of opaque bags acts as a vapor barrier as well as an electrostatic shield. However, the metallized layer in transparent bags is deposited as a grid or transparent film so that the contents will be visible. The construction of these bags is shown in FIG. 5-3. Both have interior and exterior static-dissipative surfaces.

Protective Materials and Packaging

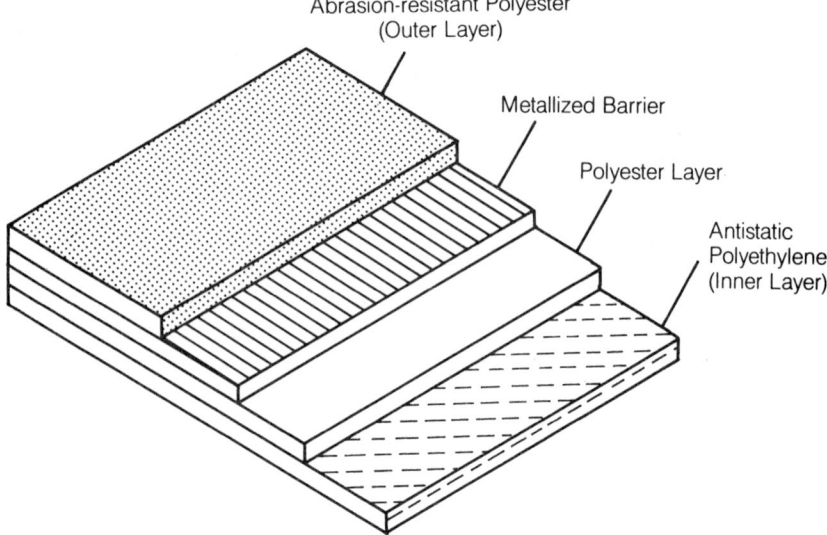

Fig. 5-3. Construction of a laminated ESD-protective bag (Faraday cage).

Transparent Faraday-cage shielding bags (see FIG. 5-4) are popular because the contents of the bag can be seen for the identification and inventory of the parts, without removing them from the bag. Verification of parts is a typical incoming inspection requirement. Transparent bags save on verification time and minimize parts handling, thus reducing the exposure of sensitive devices to possible ESD damage.

There are some differences in the specifications for transparent conductive bags, but most of the critical characteristics are similar. The outer layers are abrasion resistant, but they still permit direct measurements to be made of electrical continuity. These layers are typically made from 1/2-mil polyester with a surface resistivity of less than 10^{12} ohms/square.

The bag's transparent metallized layer is highly conductive and can dissipate charges before damage occurs. This layer is about 30 angstroms thick and has a resistivity of about 30 ohms/square. An antistatic polyethylene inner layer retards triboelectric static charging from inside the bag. This layer is typically 2.5 to 3.0 mils thick with a surface resistivity of less than 10^{12} ohms/square. Overall wall thickness of the bag varies from about 2.5 to 3.5 mils.

The static-charge decay of transparent bags is typically specified at less than 100 microseconds for a 5000-volt charge. Antistatic layers are tested to meet the requirements of MIL-B-81705 and the entire bag is expected to meet the requirements of MIL-M-38510 and EIA-541.

Some transparent static-shielding bags have laminated antistatic closures, as shown in FIG. 5-5. These closures are easy to open and close and they provide a watertight seal.

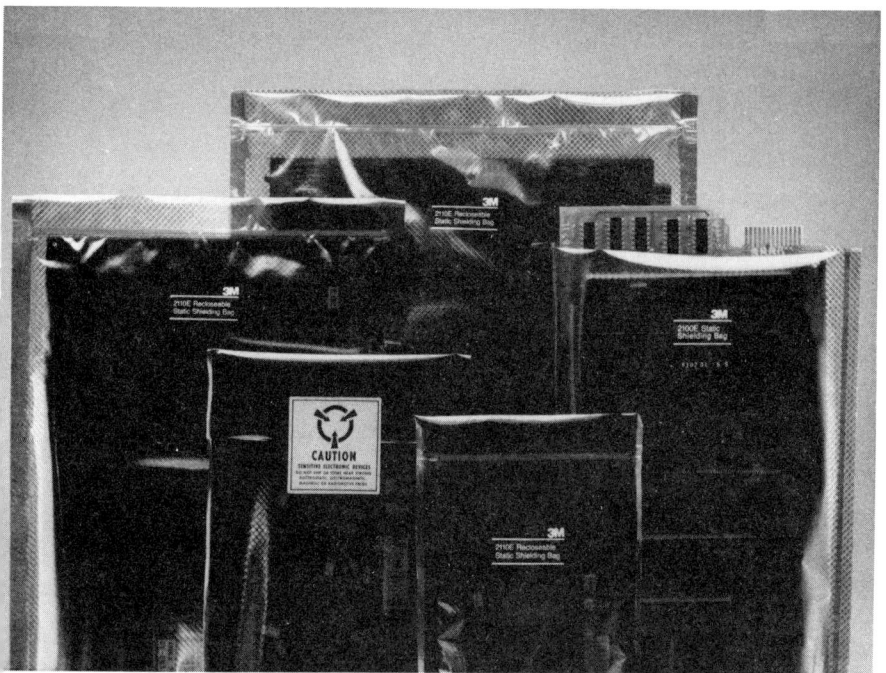

Fig. 5-4. Transparent static shielding bags.

Some transparent Faraday-cage bags are made as bubble bags. They have stainless-steel fiber film laminated between antistatic-barrier bubbles and a layer of antistatic polyethylene. The bubble cushion eliminates the need for extra padding when the bags are used with shipping cartons in the correct-matching sizes.

Low-cost, opaque, conductive plastic bags are also used to protect sensitive parts. Typical ratings are: volume resistivity less than 3000 ohm-cm and surface resistivity less than 3×10^4 ohms/square; the wall thickness is 4-mils. These bags cannot be in clean rooms because they release carbon particles if they are scratched. Also, they must be used with care because they are made of low-resistance materials that can short live circuits, causing electrical shocks and property damage.

DIP MAGAZINES

ESD-sensitive semiconductor devices packaged in DIP cases can be shipped in tubes or magazines with a C-shaped cross section, commonly referred to as "DIP sticks" (see FIG. 5-6). These packages are made from extruded aluminum or plastic that can be treated to increase conductivity.

Protective Materials and Packaging

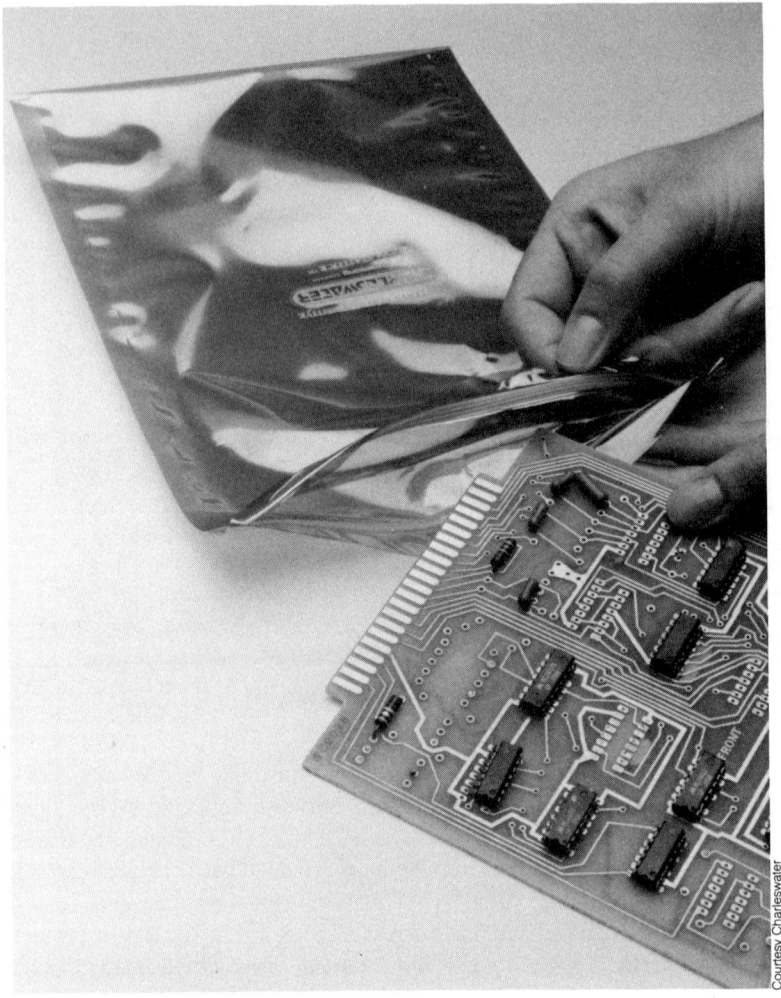

Fig. 5-5. Laminated closure on a static shielded bag.

During the manufacture of a plastic DIP-packaged IC, the silicon chip is bonded to a stamped metal lead frame and internal wire bonds are made. Then the central portion of the metal lead frame containing the chip is placed in a rectangular mold and injected with epoxy plastic. The outer lead frame is cut from the molded device, freeing the individual leads, which are then bent downward to form a configuration that will plug directly into sockets or properly aligned holes in the circuit board. The leads of ceramic DIPs are brazed to metallized ceramic package forms. The chip is bonded inside, the wire bonds are made, and the package is closed with a cover and sealed.

DIP Magazines

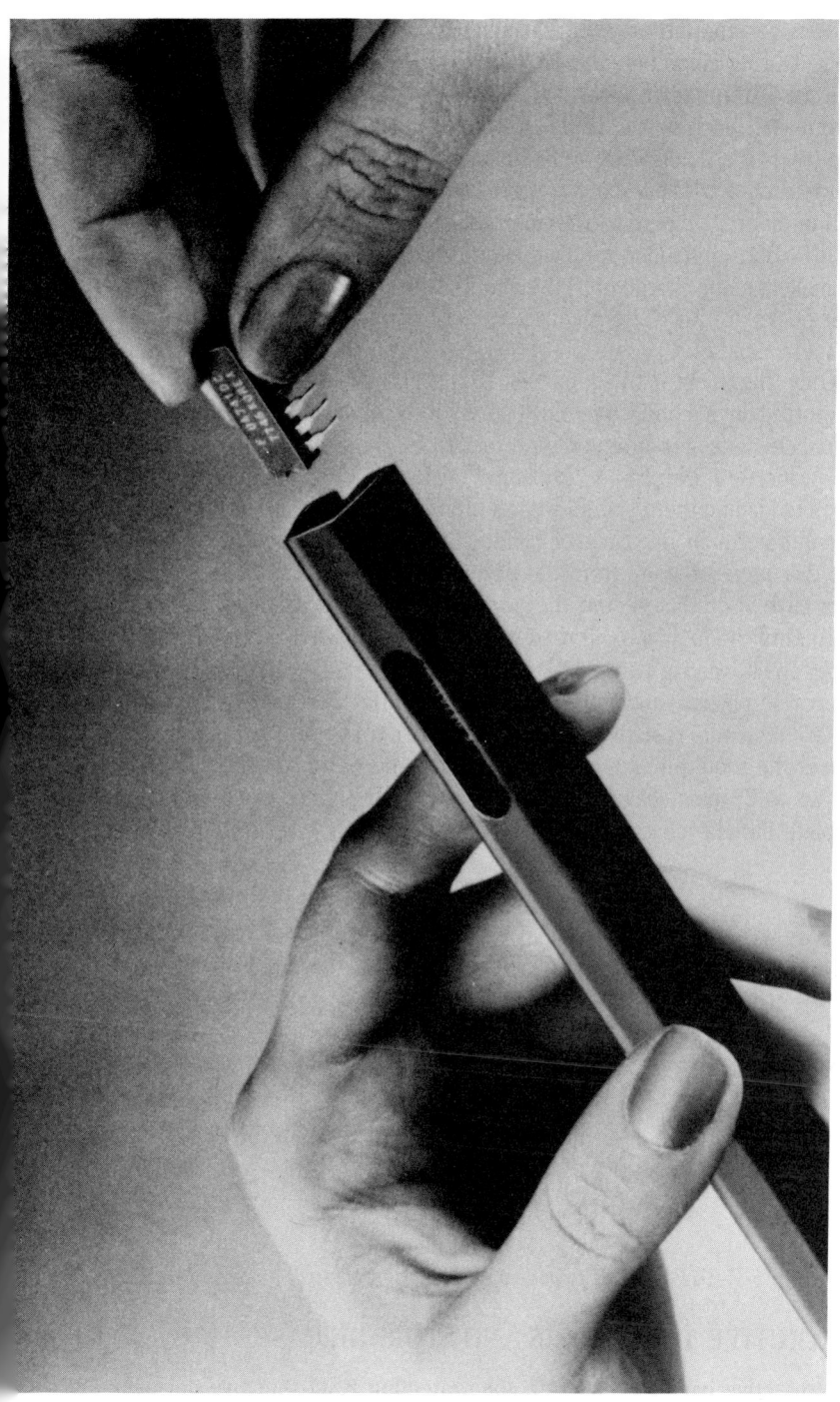

Fig. 5-6. Dual-in-line package (DIP) magazine or stick.

Protective Materials and Packaging

However, the DIP cases are insulators, and the leads are conductive metals. During shipment, the DIP-packaged devices might shift slightly inside the DIP tubes or sticks and acquire a static charge. When the DIPs are removed, each device is likely to be grounded. At that time, a rapid static discharge could destroy or damage the device. The static charge generated on a DIP-packaged device, when it slides in a DIP stick, is a result of different materials rubbing against one another.

DIP sticks, extruded from aluminum without an anodic coating, cause the smallest static charge of all the materials used to make the DIP sticks. Other sticks are made from antistatic-coated vinyl (PVC), carbon-loaded PVC, and clear PVC.

Only the most ESD-sensitive DIP-packaged devices need to be transported in a conductive DIP stick. Most sticks provide sufficient internal clearances to give some protection from magnetic fields. Faraday cage properties are claimed for some sticks. To protect the DIPs, the person unloading them should wear a grounded wrist strap. This is especially important in the case of conductive sticks because of a possible rapid discharge of static from the sliding devices is possible. Some DIP sticks have slots that permit the devices to be viewed and counted, and their identification data to be read without removing them from the DIPs.

As an alternative to the DIP stick, conductive DIP carriers (see FIG. 5-7) provide physical and static protection for DIPs. The conductive carriers have a volume resistivity of less than 500 ohm-centimeters (ohm-cm) to meet the requirements of EIA-541. These carriers, suitable for DIPs, with up to 28 pins, have hinged covers that snap shut securely, giving complete Faraday cage protection.

Fig. 5-7. DIP carriers provide both ESD and physical protection.

PROTECTIVE TOTE BOXES AND STORAGE CASES

Protective bags are essential for protecting ESD-sensitive devices, subassemblies, and circuit boards during shipment and storage. However,

Protective Tote Boxes and Storage Cases

when large quantities of devices are being transported between static-controlled workstations, it might be more convenient to carry them in covered ESD-protective tote trays or boxes (see FIG. 5-8.) Many of these containers and cases have hinged covers to provide complete ESD protection and to keep out dust during transit.

Tote boxes for circuit boards are typically molded from conductive high-density polyethylene with a typical surface resistivity of less than 3 × 10^{14} ohms-square and resistivity of less than 300 ohm-cm (see FIG. 5-9). Electrically conductive side walls and dividers permit the secure transportation of circuit boards within the tote boxes by adjusting the slots to fit the outside dimensions of the boards or cards. With snap-on covers installed, these tote boxes meet EIA-541 requirements.

Charge decay on these tote boxes should be less than 200 nano seconds for 5000 volts to reach zero volts, per mil-B-81705. With snap-on

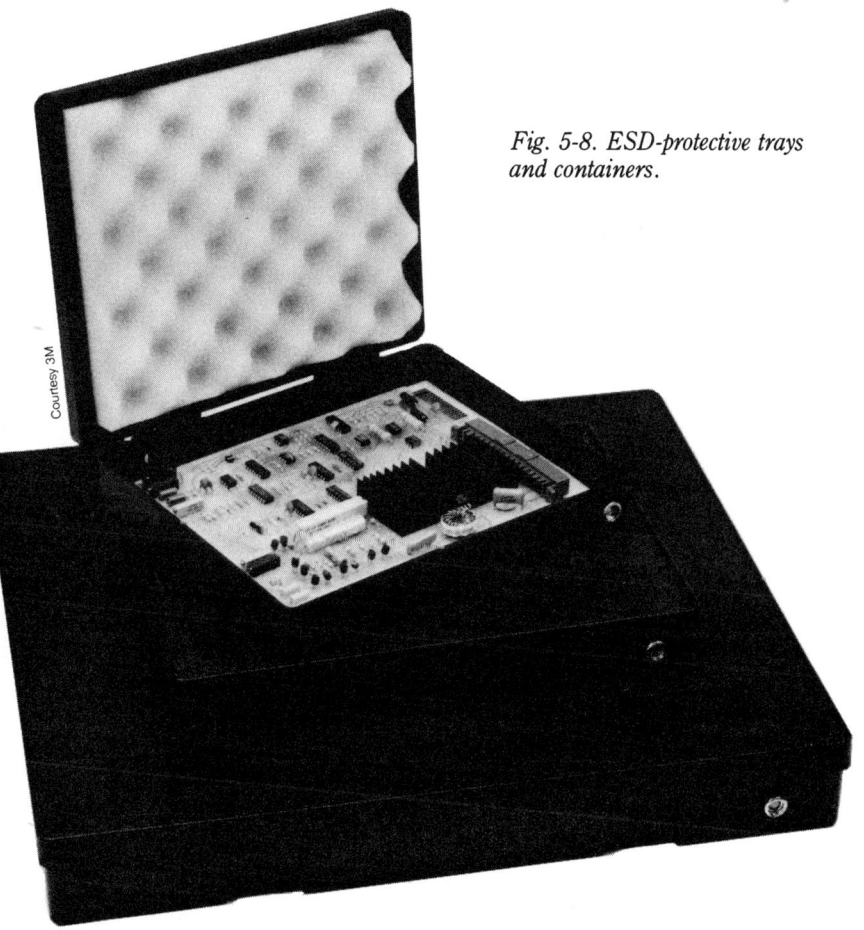

Fig. 5-8. ESD-protective trays and containers.

Protective Materials and Packaging

Fig. 5-9. ESD-protective tote boxes for circuit boards.

lids installed, these tote boxes should be able to pass the Faraday-cage test, as specified in mil-B-81075 and DOD handbook 263.

As with protective bags, higher conductivity yields higher protection against external fields. If the conductivity is too great, however, the static decay time will be too short, causing a spark discharge.

Protective Foam

Because conventional plastic sponge—usually referred to as *foam*—readily produces static charges, it should be kept away from ESD-sensitive devices. Electrically conductive polyethylene foams have been developed for protecting ESD-sensitive devices during shipment. The carbon is compounded directly into the polyethylene. The foam must be noncorrosive and unaffected by changes in relative humidity and temperature. The foams, available in vacuum-formable sheets, have different densities for different properties. Figure 5-10 shows a selection of these materials.

Foams prevent bending or misalignment of device leads, absorb vibrations and physical shocks, and short device leads to provide equipotential bonding. Low-density, two-pound conductive foam provides high conductivity and low puncture resistance to permit the safe insertion of delicate device leads. Medium-density foam (approximately three pounds) is primarily a cushioning material, but it can be used to insert sturdy electronic devices or leads. The densest foam (approximately six

Circuit Board Shunts

pounds) is primarily used for padding heavy objects, lining tote boxes, and even covering tabletops. Typical specifications for both low- and medium-density, as well as rigid, foam are a surface resistivity of less than 3×10^4 ohms/square and a volume resistivity of less than 3,000 ohm-cm.

Foams do not give complete protection to ESD-sensitive devices. For example, if a statically charged person touches the floating metal lid of a DIP, an arc can form between the lid's inside surface and the chip's top surface. This arc can destroy or damage the chip.

Fig. 5-10. Selection of conductive foam.

Protective Materials and Packaging

Fig. 5-11. Antistatic shunt bar protects circuit board edge terminals.

CIRCUIT BOARD SHUNTS

Circuit board conductive shunts, also known as *card edge protectors* or *shunt bars*, are designed to clamp over the edges of circuit boards with metallized conductors that terminate at the edges of the boards or cards. Figure 5-11 illustrates a typical antistatic shunt bar. It shorts all the conductive paths terminating on both sides of the board edges. This helps to protect ESD-sensitive components on the board or card while the board is being handled or shipped.

Typical shunt bars are made from a conductive low-density polyethylene acrylate copolymer. Volume resisitivity is typically less than 300 ohm-cm. These bars also protect the metallized sections of the circuit-board edges from physical damage during handling.

6
Protective Workplace Environment

An ESD-controlled environment is essential for any workplace where ESD-sensitive devices and circuit boards are handled for testing, inspection, sorting, assembly, and shipping. This chapter covers humidity control, work surfaces for benches or tables, flooring and floor mats, air ionizers, grounded tools, and other equipment necessary for the safe handling of these components and assemblies.

ESD control is primarily a people problem. Persons handling ESD-sensitive products should be properly grounded and wearing outer clothing made from appropriate materials, if not from static-dissipative fabric. Personal ESD-protective measures are covered in chapter 7.

HUMIDITY CONTROL

A thin layer of moisture forms on the surface of objects in rooms where the relative humidity is higher than about 50 percent. This surface film provides a conductive path to dissipate any electrostatic charge that exists or can be generated on the surfaces of walls, floors, furniture, or tools. Moisture in the air also neutralizes surface charges and adds lubricity to materials to prevent frictional charging. In general, higher relative room humidity yields lower electrostatic potential (see TABLE 6-1).

The values of electrostatic voltages measured for different kinds of human actions vary widely, particularly at low relative humidity. These variations can be traced to the many variables involved in conducting these studies: the conditions of the test, the materials involved, and even the instruments used to make the measurements. However, all the test results generally agree on the order of numbers obtained in making these measurements under various humidity conditions.

Protective Workplace Environment

Table 6-1. Effects of Humidity on Electrostatic Voltages (DOD-HDBK-263).

Means of Static Generation	Electrostatic Voltages	
	10 to 20 Percent Relative Humidity	65 to 90 Percent Relative Humidity
Walking across carpet	35,000	1,500
Walking over vinyl floor	12,000	250
Person sitting at bench	6,000	100
Opening vinyl envelope for instructions	7,000	600
Picking up polyethylene bag from bench	20,000	1,200
Sitting on chair padded with polyurethane foam	18,000	1,500

The studies have shown that adding moisture to the air dissipates electrostatic charges, and a direct relationship has been established between the amount of moisture added to the air and the reduction of electrostatic voltages. But it has also been shown that electrostatic charges are not eliminated, even at relative humidity levels above 90 percent. In addition, serious problems are created in rooms with a relative humidity of 65 to 90 percent. First, the room is very damp and uncomfortable, perhaps even intolerable for persons doing close, exacting work for up to eight hours a day. In addition, high relative humidity accelerates the rusting or oxidation of components, hardware, and tools. High humidity also establishes conditions conducive to the formation of electric leakage paths on circuit boards. Moreover, the prolonged exposure of circuit boards to moisture can cause them to delaminate during later soldering operations.

The optimum relative humidity level in ESD-controlled areas is now set by some at 40 to 50 percent, and by others at 40 to 60 percent. Special plant humidification equipment might be needed to maintain these levels of humidity, especially in northern climates where winter heating dries the room air. Humidification is usually required throughout the year in desert regions, where the outside relative humidity rarely exceeds 50 percent.

Some humidification is normally performed by the plant heating, ventilating, and air conditioning systems, but special provisions might be needed for those areas of the building where ESD-sensitive products are being handled. The installation of this equipment and its 24-hour-per-day operation is expensive.

Fortunately, air ionizers, used in conjunction with humidification, can eliminate the need for maintaining relative humidities in excess of 60 percent, and relax requirements on the regulation of that humidity. Ionization also helps to dissipate electrostatic charges.

PROTECTIVE WORK SURFACES

The surfaces of tables and benches on which ESD-sensitive components and circuits are handled should be covered with sheets of static-dissipative materials or mats. These protective work surfaces are an important defense against ESD damage. The surfaces should be capable of quickly and safely draining electrostatic charges from any conductor placed on them to ground. The discharge of conductors should be rapid to prevent damage to sensitive components, but not so rapid that a potentially destructive or damaging arc is caused. However, nonconductors placed on the mats will not be neutralized.

Wrist straps (described in detail in chapter 7) should always be worn in ESD-controlled areas to prevent static buildup on persons from discharging an ESD-sensitive device to the work surface. If wrist straps cannot be used because of the restraint in movement they impose, foot or shoe ground straps should be employed as an alternative.

Three categories of materials are used to make ESD-protective work surfaces: conductive, static dissipative, and antistatic. These categories are based on definitions of resistivity in DOD-Handbook-263: *conductive*, a material having a surface resistivity less than 10^5 ohms/square; *static dissipative*, a material having a surface resistivity between 10^5 and 10^9 ohms/square; and *antistatic*, a material having a surface resistivity between 10^9 and 10^{14} ohms/square. However, the established categories are less important than the measured values.

The characteristics of each material, as they apply to benchtops, mats, and other work surfaces, are given in TABLE 6-2. Authorities on ESD control do not agree on any precise range of resistivity values for each activity performed at the benches. In fact, most agree that there is a place in ESD control for materials in each resistivity category.

Static-dissipative materials represent a satisfactory compromise in midrange resistivity values. Antistatic surfaces generally take too long to dissipate a static charge, while conductive surfaces generally dissipate static charges too rapidly, posing the threat of arcing.

Work surfaces should be connected to a *soft ground*; that is, a conductive connection to ground in series with a resistor of a specified rating. This resistor should be located at or near the point of contact with the work surface. It should have a large enough value to limit any leakage current to 5 milliamperes or less. Moreover, the resistor should take into account the highest voltage of any source within reach of a grounded person and all parallel resistances to ground, including wrist straps, tabletops, and conductive floors. Typically this resistance value is 1 megohm.

Figure 6-1 shows a typical ESD-controlled workstation including a grounded workbench. Suitable work surfaces on the bench can either be soft or hard, but they should be appropriate to the work being performed. It is important that all bench surface materials are durable and resistant to

Protective Workplace Environment

Table 6-2. ESD-Protective Work Surface Characteristics (DOD-HDBK-263).

Conductive	Static Dissipative	Antistatic
1. Dissipates charges rapidly throughout the material and to ground, and will not maintain a high static voltage.	1. Charge dissipation rate generally adequate for most ESDS parts.	1. Provides slow bleed-off of static charges if ground straps are used by personnel working at the work bench. High ESD voltages should be rapidly dissipated through the ground strap.
2. Could discharge an ESD in the form of a spark causing EMI.	2. Provides greater resistance for personnel protection from high voltages or hard grounding if the tabletop is contacted with test equipment ground.	2. Eliminates sparks from ESD.
3. Could cause a high current discharge through an ESD-sensitive part.	3. Reduces discharge currents through ESD-sensitive part.	3. Limits discharge currents through ESDS parts to low levels.
4. Could present a safety hazard or short if a high voltage source contacted the benchtop. Could hard ground the tabletop if test equipment with grounded chassis contacted the benchtop surface.	4. Safety could require that series resistances be provided in connection to ground where high voltages can be contacted by personnel.	4. Generally provides adequate resistance for personnel safety.
5. Safety could require that series resistances be provided in connection to ground where high voltages can be contacted by personnel.		

hot solder, fluxes, and chemicals. It is also important that they are certified nonflammable in compliance with recognized safety specifications. They should also be able to withstand regular cleaning with detergents.

A typical soft work surface is a nonflammable conductive vinyl mat. Static dissipative materials with resistivities between 10^5 and 10^9 ohms/square are suitable. Alternative harder surface materials include cross-linked polymeric with surface resistivities of 10^8 to 10^9 ohms/square and fiber-reinforced polyester with surface resistivities between 10^5 and 10^8 ohms/square. Static decay should be less than 50 milliseconds.

An example of a work surface made for ESD-controlled areas is a three-layer dissipative mat from Static Inc. An antistatic top layer has a resistivity of less than 10^9 ohms/square. This surface is said to move

Protective Work Surfaces

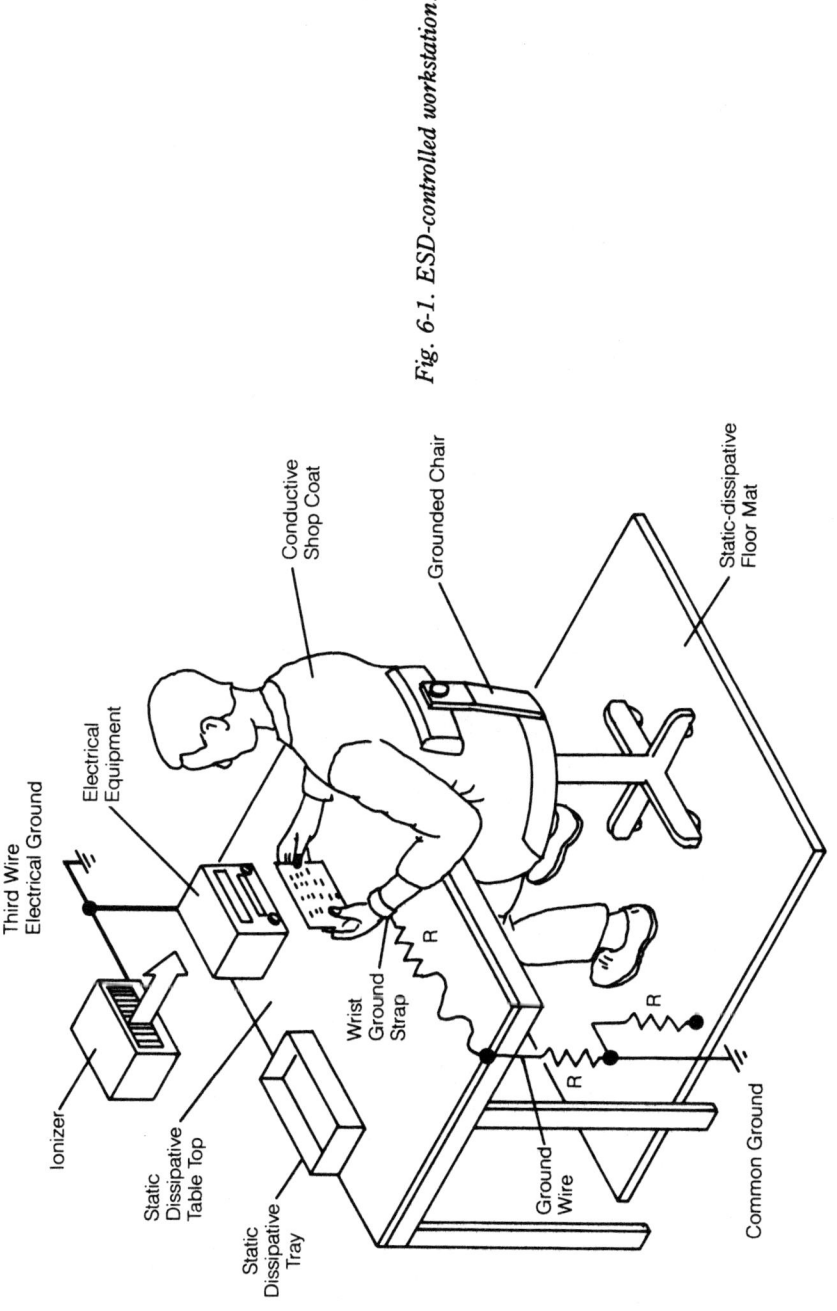

Fig. 6-1. ESD-controlled workstation.

Protective Workplace Environment

potentially damaging static charges through to the highly conductive continuous middle layer, with a volume resistivity of 150 ohm-cm serving as the main static drain. A foamed vinyl antistatic bottom layer also has surface resistivity of less than 10^9 ohms/square; it provides cushioning and skid resistance. The static decay of the top layer is less than 0.04 seconds and its surface resistance, measured as a laminate, is less than 4×10^6 ohms.

Some work surface materials are available as mats measuring up to 30×40 inches with widths from 24 to 54 inches (see FIG. 6-2). Fasteners or receptacles attached to the material permit the connection of ground cords, terminated by heavy-duty clips or banana-plug receptacles.

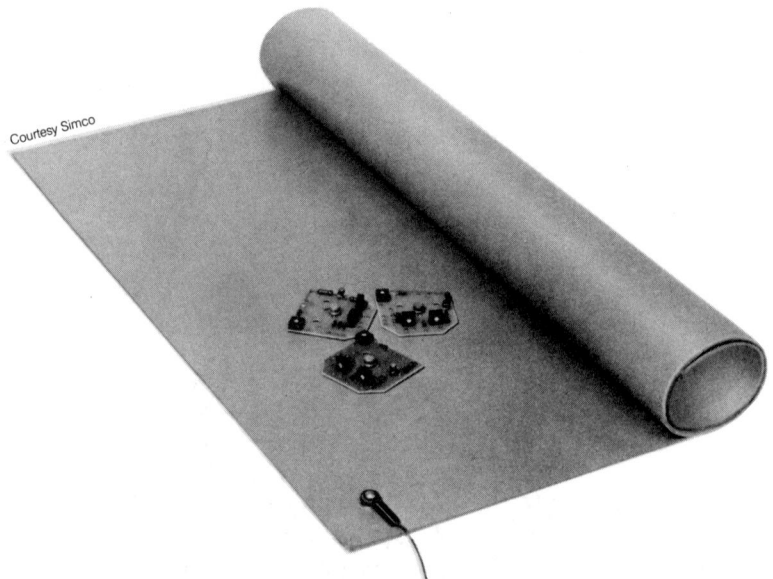

Fig. 6-2. Static dissipative table mat material.

Electric Shock Hazard

Precautionary steps must be taken when furnishing ESD-controlled work areas and grounded workbenches to reduce the possibility of electrical shock to personnel. Before these steps are considered, it is useful to review the physiological effects of electrical shock.

The severity of electrical shock is determined by the magnitude and path of the current flowing through the body and the duration of current flow. TABLE 6-3 shows that even relatively small currents can be fatal to humans if the conductive path includes a vital part of the body, such as the heart or lungs. The voltage necessary to produce a fatal current depends on the resistance of the body, contact conditions, and the path through the

Protective Work Surfaces

Table 6-3. Effects of Electrical Current on the Human Body (DOD-HDBK-263).

Current Values (Milliamperes)		Effects
Ac 60 Hz	Dc	
0 – 1	0 – 4	Perception
1 – 4	4 – 15	Surprise
4 – 21	15 – 80	Reflex action
21 – 40	80 – 160	Muscular inhibition
40 – 100	160 – 300	Respiratory block
Over 100	Over 300	Usually fatal

body. In addition to the danger of shock, there is also the hazard of electrical burns produced by the heat from the arc that occurs when a part of the body touches a high-voltage circuit. The burns are caused by the current passing through the skin and tissues.

Grounding Electronic Test Equipment and Tools

Because of shock hazards in protected areas and on grounded workbenches, all external parts, surfaces, and shields in electronic test equipment and power tools must always be at a common ground potential during normal operation. Where a ground is part of the circuit, the external or interconnecting cable should carry a ground wire in the cable, terminated at both ends, in the same manner as the other conductors.

With the exception of coaxial cable shields, metallic covering on or in other cables is not as dependable as current-carrying ground connections. Plugs and convenience outlets, for use with metal-encased portable power tools and equipment, should ground the metal frame or case, and equipment when the plug is mated with the receptacle; the grounding pin should make first and break last.

Extreme caution should be exercised in placing these tools and test equipment on grounded workbenches with metal or other conductive covering. The hard grounded cases of the tools and test equipment can shunt the protective resistance in the workbench ground cable.

For further protection of personnel, ground fault interrupters (GFIs) should be used with test equipment. The GFI senses leakage current from faulty test equipment and instantly interrupts the circuit when the current reaches a potentially hazardous level.

It is important to avoid parallel paths to ground. These paths could reduce the equivalent resistance of persons to ground to below safe levels. Parallel paths could result from the use of wrist straps, grounded tabletops, and grounded floor mats.

Protective Workplace Environment

Ground Potential of Grounded Work Surfaces

Additional safety and grounding considerations for ESD-protected areas and grounded work surfaces are:

- Cables and resistors should have ample current-carrying capacity. Because the work surface ground is intended to bleed off electrostatic charges, a $1/2$-watt resistor is usually used.
- Ground cable connections should be continuous and permanent.
- The ground cable and connection should be made from a material that is strong enough to minimize accidental ground disconnections.
- Workstation mats and surfaces, floor mats, ground straps, and other ESD-protected area grounds used to discharge static electricity should be connected to earth, the power system, or other hard ground as appropriate through current-limiting resistances. The wrist strap should be connected to ground through the ground point of the workbench top (see FIG. 6-1). Workbenches should not be connected in series with one another because the series resistances will add together, giving a higher ESD dissipation time. Also, an opening in one ground cable could affect the other workbench ground cables.

Safety considerations specify the minimum resistance to ground for protection of personnel. The maximum resistance to hard ground for personnel grounding is determined by the decay time for an electrostatic charge. This decay time is determined by human capacitance, and resistance, and the resistance of other ground paths to hard ground. This time should be short enough to dissipate charges at or below the rate at which they are normally generated.

Durability Considerations

In addition to electrical conduction considerations, several other physical properties must be evaluated during the selection of materials for work surfaces. If the surface is too hard, components and circuit boards can be damaged from being dropped. But if the surface is too soft, it is subject to abrasion, scoring, and rapid wear; also this process might release undesirable particulates in the work area. The surface should be resistant to chemical solvents such as trichlorethylene that are frequently employed in electronics assembly and test procedures.

The activity being performed at the workstation might dictate the properties of the work surface and their importance. For example, a work surface for the repair of circuit boards might be too soft and therefore unsuitable for mechanical assembly.

ESD-protective Floors and Finishes

Most conventional floor surfaces—including painted or sealed concrete; shellacked, varnished, or painted wood; vinyl tile; or carpet—can contribute significantly to the factory ESD problem. Consequently, these floor surfaces should either be replaced or covered with ESD-protective flooring or floor mats.

Protective floor surfaces, like ESD-protective bench and table work surfaces, are available either as conductive, static-dissipative, or antistatic materials. The characteristics of each of these materials are given in TABLE 6-4. These materials are available in the form of floor mats, vinyl floor tiles, and terrazzo pieces.

Conductive rubber mats provide a low-cost, effective static control surface for general workstation use. This material is able to withstand hot solder and solvents. Vinyl floor mats are durable and provide comfortable surfaces.

Conductive vinyl floor tiles have long been used in hospital operating rooms and intensive care stations to prevent explosions of flammable air and gas mixtures caused by ESD. Its permanent static control properties are achieved by mixing conductive elements in a matrix of flexible vinyl resin. This tile is made to conduct static charges to ground through conduction from tile to tile. The surface resistivity of this tile should be in the range of 10^4 to 10^6 ohms/square, and it should permit the decay of $\pm 5,000$ volts to 0 in less than 50 milliseconds.

One type of semipermanent conductive flooring now available is made of flexible, interlocking modular tiles. Designed for grounding moving personnel in either temporary or permanent installations, its open grid tiles

Table 6-4. ESD-Protective Floor Mats Characteristics (DOD-HDBK-263).

Conductive	Static Dissipative	Antistatic
1. Dissipates charges rapidly throughout the material and to ground, and will not maintain a high static voltage.	1. Provides adequate conductivity for dissipation of charges.	1. Provides slow bleed-off of high static charges.
2. Safety could require that series resistances be provided in connection to ground where high voltages can be contacted by personnel.	2. Generally provides sufficient resistance for personnel safety. External series resistance to ground may not be required.	2. Accumulations of dirt, contaminants and wear reduce antistatic properties. Requires frequent cleaning and treatment with a topical antistat.

Protective Workplace Environment

are maintenance-free, skid-proof, and resistant to most solvents. Because these tiles are modular and interlocking, they can fit to any floor size. As manufacturing floor area requirements change, tiles can be added or removed. The interlocking tabs permit electrical continuity to be maintained at all times. The typical surface resistivity of this kind of flooring is 1.7×10^5 ohms/square.

Floor mats are another alternative for ESD-preventative floor covering. Figure 6-3 illustrates a black-textured vinyl antistatic floor mat; FIG. 6-4, a transparent conductive floor mat. The typical specifications for a vinyl mat are: surface resistivity of less than 10^9 ohms/square, and resistance-to-ground of less than 10^9 ohms. It should dissipate a 5,000-volt charge in less than 50 milliseconds.

Although convenient and easy to install, conductive floor mats that are not fastened to the floor and are not carefully placed outside of traffic lanes, can constitute a tripping hazard. Moreover, the mats must be

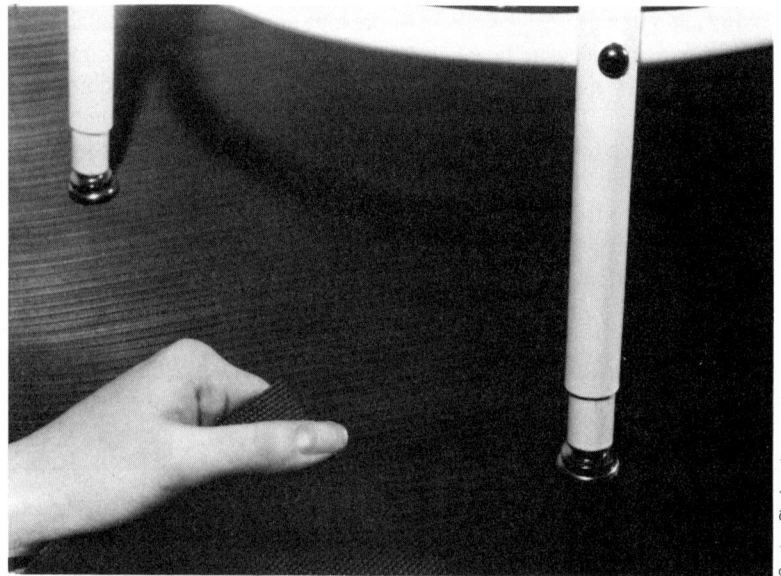

Fig. 6-3. Black textured vinyl antistatic chair mat.

moved for cleaning; this may release dust and contaminants into the air. Just as with ESD-protective work surfaces, conductive floors or floor mats must have a current-limiting resistor in the ground lead.

When conductive floors or mats are used, personnel should wear conductive shoes, shoe covers, or heel grounders because shoes made of materials other than leather have inadequate conductivity. Shoe grounders are discussed in detail in chapter 7.

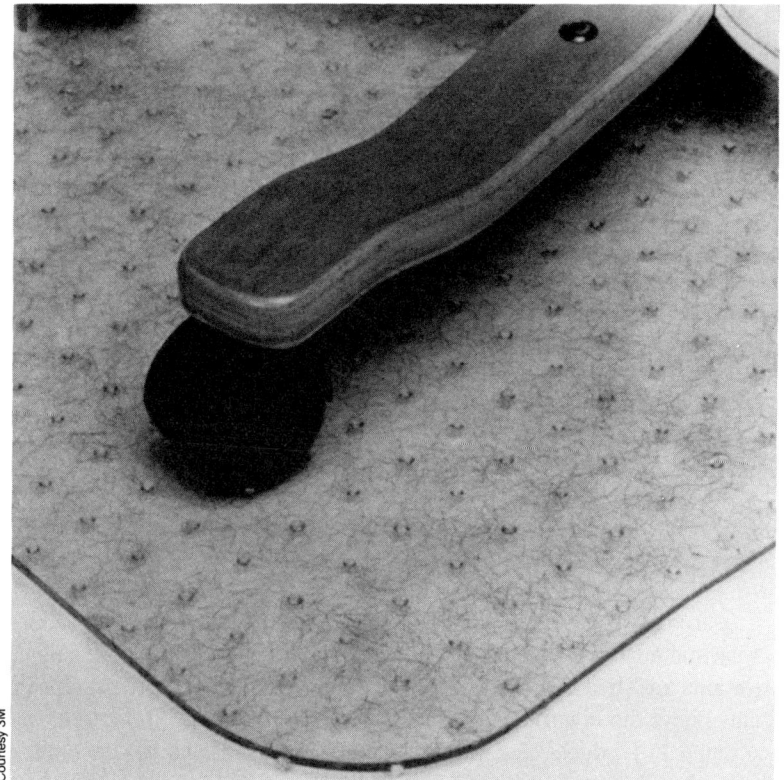

Fig. 6-4. Transparent conductive mat dissipates static charges.

Shoes, shoe covers, and ground straps must be kept clean at all times, or they will become contaminated by dirt or chemicals and lose their conductivity to the floors or mats. Because of the hazard of sliding, ground straps should not be worn outside an ESD-protected area.

Conductive floor finishes can be mopped or painted onto floors to provide ESD protection. Because of foot and cart traffic, however, these finishes must be reapplied periodically. Regardless of the protection used, protective floors must not be waxed because wax buildup will reduce the conductivity of the floor.

Suitable high-gloss industrial acrylic floor finishes with proprietary conductive agents are available as commercial products. This finish should have a surface resistivity of 10^6 to 10^9 ohms/square, in accordance with ASTM D-257, and it should be able to dissipate a 5,000-volt charge to 0 volts in less than 100 milliseconds, in accordance with MIL-B-81705 or Federal Test Method Standard 101B, Method 4046.

Suitable conductive floor paint should have a surface resistivity of 10^4 to 10^6 ohms/square. In addition to floors, this paint can be used on tote boxes, utility containers, walls, and wood or plastic furniture.

Protective Workplace Environment

ESD-conductive Seating

Conductive seat cushions are required on stools or chairs used in the ESD-protected area since a person sitting at a workbench or table might lift his or her feet from the conductive floor to the rungs of the stool or chair. If the stool or chair is not grounded, the protective qualities of the flooring will be lost. Some chair cushions are upholstered with a static-dissipative nylon fabric connected to a brass drag chain to provide a dissipative path to the grounded chair mat.

Dissipative chairs and stools are made for use in ESD-protected rooms. When used with conductive floor mats, electrostatic charges on personnel as high as 18,000 volts are discharged to 0 volts in less than 0.1 second, in accordance with tests specified in DOD-HDBK-263.

AIR IONIZERS

As stated earlier, there is no effective way to ground nonconductive items. For this reasons, air ionizers have become important for neutralizing static charges. However, ionizers do not eliminate the need for either wrist straps or conductive work surfaces.

Air ionizers generate positive and negative air ions. The positive ions are attracted to negatively charged nonconductive surfaces, and the negative ions are attracted to positively charged surfaces. The attraction continues until the electric field is neutralized. At that time, the ionizer will continue to produce ions; the unneeded or unused ions will recombine with other ions circulating in the air.

Air ionization occurs when the air receives sufficient energy from induced charge, high-voltage ac or dc, or radioactive emissions. When this occurs, an electron is dislodged from an atom or molecule, leaving the atom or molecule with a net positive charge. The free electron will soon be attracted to a neutral air molecule, forming a negative ion. The atom or molecule that lost the electron becomes a positive ion.

The output of air ionizers should contain nearly equal quantities of positive and negative ions to dissipate both the negative and positive charges produced when static electricity is generated. If an imbalance of positive and negative ions exists, residual voltages can develop over the ionized area. Three types of air ionizers are commonly used: static comb, electric-powered (see FIG. 6-5), and radioactive or nuclear.

The static comb ionizer is a simple, inexpensive unit that requires no power supply. It consists of either grounded needles or metallic brushes placed near the surface to be neutralized. The charged surface creates a potential gradient between the grounded needles and itself. When the voltage becomes high enough, the air will ionize and create a conductive path to ground through the needles. Unfortunately, a charge of at least 2,500 volts is required for the static comb ionizer to work.

Air Ionizers

Fig. 6-5. Electric-powered air ionizer with heater and fan.

Electrostatic voltages far less than 2,500 volts can damage or destroy semiconductor devices. Therefore, these passive ionizers are primarily used to reduce the effects of static charges in plants where strips of plastic or paper are moving at high speeds. The static comb ionizer rarely finds application in electronics plants.

Electric-powered ionizers consist of two basic components: a high-voltage power supply and an ion producing section, generally organized as a series of needles mounted on a bar. The high-voltage power source can supply either dc or ac current to the needles. Ac will alternately produce positive and negative ions. Ac ionizers can use conventional transformers or solid-state circuity.

Transformer power supplies might produce voltages as high as 20,000 volts, generally at a frequency of 60 Hz. Because ionization occurs only at the peaks of these waveforms, there are periods of several microseconds between the peaks when no ions are being produced.

Because of the high voltages present, ozone can be an undesirable by-product of an air ionizer. Fortunately, ozone production from air ionizers approved for use in ESD-protected rooms and spaces is not significant. According to the Occupational Safety and Health Act, ventilation in an enclosed work space should be sufficient to maintain the atmospheric

Protective Workplace Environment

concentration of ozone below 0.1 part per million, the maximum level allowed.

Solid-state power supplies avoid significant ozone generation because they operate at lower voltages—3,000 to 4,000 volts. Power-supply operation at reduced levels is possible because the frequency, amplitude, and pulse width can be controlled. The operating frequency of these supplies is between 15 and 40 kHz, in contrast with the 60 or 120 Hz produced by conventional linear power supplies. Because of their higher frequency, lower voltage, and pulse width control, these power supplies also produce much less electromagnetic interference (EMI) than linear power supplies.

The output of the power supply is applied to a group of evenly spaced needles held rigidly less than 1 inch from a grounded metal housing. The air is ionized by passing it across an array of sharp needles (electrodes) that operate at a potential of thousands of volts. The sharp points are needed to create a sufficient potential gradient to achieve ionization.

When the neutralizing needles are connected directly to the power supply, (see FIG. 6-6A), the configuration is called a *hot bar*. During operation, if any of these needles is accidentally shorted to ground, the entire bar will become disabled, and the supply voltage will drop instantly.

Hot bars present an additional shock hazard. A person could accidentally touch one of the needles and receive an electric shock. The severity of the shock received depends on the rating of the power supply. If the power supply is designed to draw no more than 5 milliamperes, the person who contacts them will receive only an unpleasant sting. However, if the power supply is not limited, the shock could be fatal, depending on the current available and the effective resistance of the person. Because of

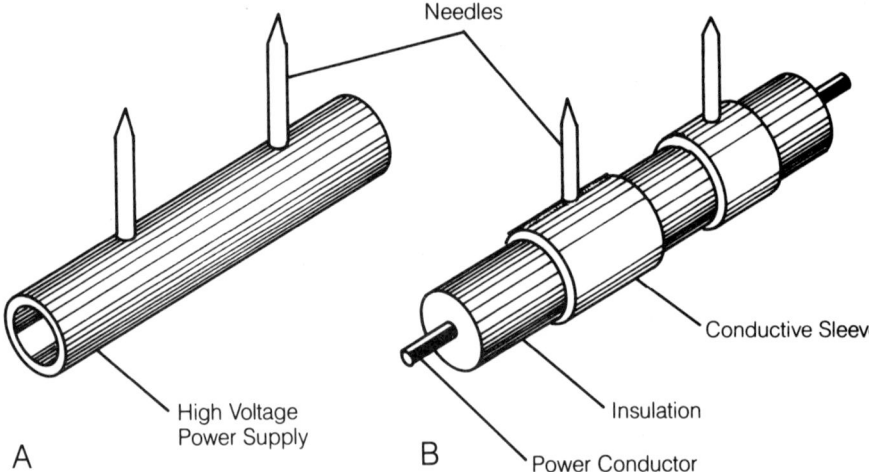

Fig. 6-6. High-voltage coupling of powered static eliminators in air ionizers.

Air Ionizers

this drawback, hot bars are usually shielded so that they cannot be touched without disabling the equipment. Where they can be used safely, hot bars are more efficient and less expensive than shockless bars.

Cold (shockless) *bars* have needles that are capacitively coupled to the power supply. Each needle is embedded in a conductive sleeve surrounding the high-voltage transmission cable (see FIG. 6-6B). In this configuration, the insulator is the dielectric of a capacitor, the cable conductor is one plate of the capacitor, and the conductive sleeve is the other plate.

Because each needle is an independent circuit, one or more needles in the assembly can be grounded without affecting the others. Typically, a short-circuit current of less than 40 microamperes is available at any point, so accidental contact will not result in a painful shock. The sharper the needle points, the more efficient the ionizer will be.

Operated at this high potential, the needles act in a manner similar to other charged surfaces: they attract oppositely charged particulate matter. The particles that accumulate on the needles might blow off and contaminate the ionized air stream, thus reducing the efficiency of the equipment. Therefore, it is necessary to keep the needles clean with regular maintenance. In addition to accumulating dust, the needles might become eroded or carbonized, reducing efficiency by up to 15 percent per year.

In many factory situations, the ionizers can be located several inches to several feet from a charged object. Because of this distance, fans are added to the ionizers to circulate the ionized air and direct a flow onto charged objects. High rates of air flow are important for rapid neutralization of the electric fields. If the fans are too powerful, however, they cause discomfort for nearby employees and can interfere with work by blowing light objects off the workbench. The noise from the fan might also be a source of annoyance for nearby personnel. Most ionized air blowers employ tubeaxial fans containing a heating element to warm the air so that cold air is not blown over personnel working within the fan's throw.

A finite amount of time is required to dissipate static charges. The magnitude of the charge and the distance between the charged surface and the ionizing source are factors in dissipation time. The time lapse can vary from several seconds to several minutes. For this reason, ionizers should be turned on in the work area at least two to three minutes before any containers of ESD-sensitive components or circuit boards are opened.

Some ionizers do not provide a balance between the positive and negative ions emitted. They might leave residual voltages high enough to damage some ESD-sensitive devices. During selection and location of ionizers, it is important that the residual voltages in the area are measured and compared with the voltage sensitivity levels of the components to be handled. Ionizers do not give permanent protection for static-producing materials. Once these materials are removed from the ionizer's field, they will again produce electrostatic charges.

In addition to producing ozone, ionizers present other problems: they might cause inadvertent high-voltage breakdown in the vicinity of high-voltage equipment; the ionizer section or the fan motor could cause a fire or explosion in rooms or spaces containing volatile air mixtures; the ionizer fan motor might cause EMI and annoying audible noise.

An example of a bench or wall-mountable ionizer is offered by Charleswater Products. This non-nuclear unit, the CP922 shown in FIG. 6-5, is said to produce less than 0.01 parts per million of ozone because of its relatively low ion source voltage. The ion source has emitters formed from an alloy of three metals. The CP922 features self-balanced ion output permitting a 1:1 positive/negative ion ratio to be maintained.

Air delivery from a tubeaxial fan is adjustable to 117 cubic feet per minute with a maximum air velocity of 450 feet per minute. The *field*, or throw, is approximately 10 feet long and 2 feet wide. Specified EOS/ESD-S3 charge decay time is from 0.9 second at 1 foot to 9.2 seconds at 5 feet. The enclosed unit has outside dimensions of $11 \times 7 \times 9$ inches. An automatic heater used with the variable speed fan keeps personnel comfortable. The heater has an on/off control and is preset 8.5 degrees above ambient temperature; it draws 330 watts. Maximum audible noise at 3 feet is given as 50 dB. The standard unit operates on 120 Vac, but a 220-Vac model is available.

NUCLEAR IONIZERS

Nuclear static eliminators or radioactive ionizers usually employ the radioisotope polonium-210 in a small cartridge to neutralize static charges. This radioisotope emits only *alpha particles*, charged helium nuclei consisting of two protons and two neutrons. The positively charged alpha particles are emitted at high velocity to pass through the air and collide with air molecules, creating positive and negative ions. Radium and americum-241 also emit alpha particles. They are not used for air ionization in populated areas because they also emit more penetrating gamma rays.

The alpha particles emitted by these radioisotopes have high energy, but a low penetrating ability. In the air at atmospheric pressure, harmless alpha particles have a range of approximately 2 inches; they can be blocked by a thin sheet of paper or the epidermal layer of skin. However, no alpha particles are present in the stream of ionized air. The particles are emitted inside a closed chamber, where they collide with air molecules.

The nuclear static eliminator overcomes some of the drawbacks of the electrical ionizer because it does not require a power supply to ionize the air. Without a fan, it does not need external power or wiring that would limit its portability. If it does not have a fan, it can be used safely in locations where there are explosion or fire hazards. However, most nuclear-powered ionizers for volume static elimination incorporate a fan as well as a heater for the comfort of personnel in the vicinity.

Unfortunately, the nuclear static eliminator has been the victim of public reaction against all nuclear or radioactive sources, no matter how harmless they actually are. Many employees prefer not to work in the presence of any radioactive source. However, no documented cases of harmful effects to humans are traceable to these air ionizers.

The polonium-210 within the capsule, like other radioactive sources, loses its effectiveness with time because the radioisotope inside decays. Radioactive $84\ Po^{210}$ emits alpha particles with an energy of 5.3 million electron volts (MeV), and it has a half-life of 138.4 days. This means that the normal useful life of a polonium-210 capsule is 1 year. At the end of that time, the nuclear element must be replaced by the manufacturer of the nuclear ionizer. To simplify this procedure, the ionizers are leased, rather than sold, by the manufacturer. The complete unit is then returned to the manufacturer, and the ionizer is replaced with another.

Ionizing Air Guns

Dry compressed air and nitrogen are widely used to remove dust, metal, and plastic chips from products being manufactured. If these objects have a static charge, however, they will attract dust once the compressed air is removed. Moreover, the moving air stream might cause buildup of additional charges on the products. Nuclear cartridges containing polonium-210 also can be included in *ionized air guns*—hand-held, high-pressure units that lift particulate matter from surfaces while they neutralize the electrostatic charge. The gun is connected to an air compressor by several feet of shielded cable.

A 1-micron pore-size membrane filter in the gun traps dust particles larger than 1 micron, and oil and moisture from the compressed air. The filter prevents the residue from being propelled with the ionized air. Ionization in hand-held ionized air guns also can be provided by high-voltage electrical discharge.

An ESD ionizer gun must be insulated against electrical shock. The guns are light weight to reduce operator fatigue with prolonged use. One manufacturer includes an ultrasonic generator to help remove small particles. The ionizing air nozzle, a variation of the ionizing air gun, supplies a narrow stream of ionized air at a fixed location. The gun could be used over a materials-handling moving belt to provide a continuous ionized air curtain, although this is not widely specified for ESD control in electronics plants.

Protective Workplace Environment

TOOLS AND PRODUCTION EQUIPMENT

Conventional metal and plastic hand tools and equipment used at the workbench can present an additional ESD threat to sensitive components. These tools—used for activities ranging from the adjustment and straightening of components to the assembly, soldering, and testing of components and circuit boards—include tweezers, scissors, pliers, screwdrivers, soldering guns, and multimeters.

Tools or probes that are made of metal and are hand-held concentrate the discharge at a point, thus increasing its potential for damaging or destroying an ESD-sensitive device. ESD currents from hand-held tools might surpass 60 amperes—over four times more than those from a finger. It is this finding that has caused ESD authorities to question the validity of the MIL-STD-883 test waveform, shown in chapter 2.

Tools made of plastic or with plastic handles can themselves act as sources of electrostatic discharge. The recognition of this fact has led to tool modifications for work on ESD-sensitive devices. Alternatively, they can be treated with antistats to eliminate ESD.

Periodically, insulated tool handles should be checked for static generation and treated with an antistat, if necessary (see the section, "Topical Antistats," in this chapter). Small tools that are frequently handled often accumulate skin moisture and oils, making them ESD protective without the addition of antistats. Nevertheless, the tools should be checked frequently, particularly if they are in contact with solvents that could remove these oils.

Powered tools, such as soldering irons, soldering guns, solder pots, and flow soldering equipment, should be hard-grounded and isolated from the power line by transformers or powered by dc. To keep the voltage buildup on a hot soldering iron under 15 volts, the resistance between the tip of the iron and ground should be less than 20 ohms. Other electrical power equipment that could contact ESD-sensitive products should also be grounded. ESD-protective solder wicks should be used.

Grounded baffles are needed in temperature chambers to dissipate charges in circulated air. However, as an alternative to baffles, ionized air can be used in the chamber to dissipate static charges caused by air flow. Also, shields can be used to divert the charged air from ESD-sensitive objects in the chamber. Carbon dioxide can be used in cooling chambers, but caution is necessary because the evaporation of carbon dioxide can generate high static charges. All components tested in temperature chambers should be placed in ESD-protective tote boxes or trays on grounded metal racks within the chamber.

When ESD-sensitive objects are sprayed, cleaned, painted, or sandblasted, ionized air blowers, conductive solvents, or ionized nozzles should be used where appropriate to prevent static charge buildup in the work area. Where practical, a wet blast conductive or antistatically treated slurry with a maximum volume resistivity of 500 ohms-cm should be used, rather than dry sandblasting. Low-resistivity solvents, such as ethanol, mixed with a normal cleaning solvent give improved control of charge generation.

TOPICAL ANTISTATS

The use of antistatic chemicals (antistats) to control the generation of static charges is optional and might supplement an organized ESD control program. Topical antistats can be applied to the surface of nonconductive or nonstatic-dissipative materials to reduce or eliminate the generation of static charges in manufacturing or storage areas. Static elimination is accomplished by increasing surface lubricity, thereby reducing the material's coefficient of friction, with less friction on the surface, triboelectric charging is reduced. In addition, topical antistats increase surface conductivity, allowing charges to bleed off.

Charge dissipation from surface conductivity can be explained by two different theories. According to one theory, increased conductivity allows the antistat and the material it covers to exchange enough electrons to maintain an electron balance. According to the second theory, however, the antistat introduces a balance of positive and negative ions to the material's surface. Enough ions are introduced to neutralize most of any charge generated on the surface. In addition, an ion exchange with the surrounding air further dissipates the charge.

Topical antistats are usually liquids consisting of a vehicle and an antistat. The vehicle—usually water, alcohol, or mineral spirits—transports the antistat to the surface of the object to be protected. The antistat remains as a deposit on the insulating material's surface after the vehicle evaporates; it is the substance that controls the static.

Some antistats are detergents that combine with the moisture in the air to wet the surfaces where they are deposited. These hygroscopic antistats are less effective in low relative humidity because less available water vapor is in the air.

Topical antistats can be applied by brushing, spraying, rolling, dipping, mopping, or wiping. Antistatic solutions must be free of reactive elements, such as chlorine and phosphorous. Antistats are particularly useful in treating soft material surfaces, including carpet, fabric seats, work-related clothing, and foam padding. Topical antistats can also be used to treat hard materials subject to abuse, such as floors, tabletops, and tools, as well as hard surfaces that receive very little abuse, such as cabinets, walls, and fixtures.

Protective Workplace Environment

Antistatic sprays or solutions should never be applied to electrical components and machines, electronic components, circuit boards, or subassemblies because the chemicals can cause leakage paths and short circuits. Moreover, antistats contaminate prepared metal surfaces, preventing molten solder from adhering to them. Antistatic chemical applications should never be performed where components, packages, or personnel are exposed to spray mists and evaporation vapors.

Antistat treatments can be removed unintentionally through contact with hands, tools, clothing, or shoes; it can be removed by routine cleaning. The need for an initial application and the frequency of reapplication can be established only through routine electrostatic voltage measurements using appropriate test instruments. The applications' effectiveness can be checked periodically by either rubbing the treated area with common polyethylene and monitoring its charge and decay time with an electrostatic fieldmeter, or by measuring the surface resistivity of a sample of the material with the appropriate test instrument (see chapter 8).

Antistats should be reapplied to carpet, fabric seats, cushions, and other soft surfaces every six months or after cleaning. Hard surfaces such as floors, tabletops, and tools should be retreated at least once a week, and after cleaning, by wiping or mopping. In high-activity areas, daily retreatment is recommended. Antistats should be reapplied to cabinets, walls, and fixtures every six or twelve months after cleaning, by wiping or spraying. Clothing and smocks should be retreated after each cleaning by spraying or adding antistatic concentrate to final rinse water while washing.

Objects that require periodic retreatment with a topical antistat should be labeled with a sticker listing the dates for checking and retreating. Tote boxes and trays should typically be labeled. Many different characteristics must be considered when selecting an antistat, aside from its antistatic properties:

- **Bacteriostatic**—will inhibit bacterial growth
- **Nontoxic**—will not irritate the skin or eyes
- **Noncorrosive**—will not accelerate corrosion or rusting if applied to bare metals
- **Nonflammable**—will not sustain ignition
- **Nonstaining**—will not leave a visible residue
- **Longevity**—will remain effective for long periods
- **Durability**—will permit some handling without loss of effectiveness
- **Ease of application**—will apply easily to objects
- **Compatibility**—will not react with materials it contacts

ESD LABELS AND SIGNS

An important part of any ESD control program is notifying and reminding personnel that the program is in effect. ESD caution signs should be displayed prominently in work areas and warning labels should be on ESD-sensitive parts and containers. The labels should be consistent in color, use of symbols, class of parts, voltage-sensitivity identification, and appropriate instructions.

Signs posted prominently at all workstations where ESD-sensitive items are handled should contain the following information or its equivalent:

CAUTION

STATIC CAN DAMAGE COMPONENTS

Do not handle ESD-sensitive items unless a grounding wrist strap is properly worn and grounded. Do not let clothing or plain plastic materials contact or come in close proximity to ESD-sensitive items.

Labels should be attached to easily visible areas of all bags, trays, tote boxes, and other containers of ESD-sensitive items. The labels should be placed consistently on containers and packages, at a standard location, to eliminate mishandling.

7
Personal Protection against ESD

Sensitive components and circuit boards should be kept in ESD-protective packages whenever possible. The components should only be removed from these packages in an ESD-controlled environment. In addition, these sensitive products should be handled only by persons wearing approved ESD-grounding straps.

This chapter covers personal ESD protective equipment, including wrist and shoe ground straps and static-dissipative outer clothing. The field service kit is included in this chapter because it is a portable ESD-controlled environment for which the service engineer or technician is responsible.

GROUNDED WRIST STRAPS

Wrist straps are the single most important personal accessories for the protection of ESD-sensitive components and circuits. They are easy to put on, more effective than most other personal protective clothing, and inexpensive. However, wrist straps only provide protection against static charges on the body; they do not dissipate static charges on the clothes of the wearer.

By maintaining an electrical path between ground and the sensitive device handler, wrist straps eliminate or substantially reduce ESD voltages that are generated by a person. Figure 7-1 shows typical voltages and waveforms generated by an ungrounded person. By contrast, a person wearing a grounded wrist strap has his or her peak voltage typically cut to less than 11 volts. The plot clearly shows the value of proper training in

Grounded Wrist Straps

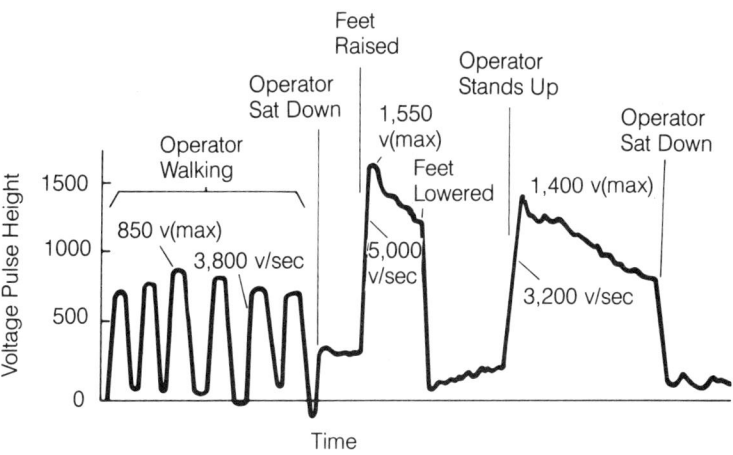

Fig. 7-1. Typical voltage variations monitored on a person not wearing a grounded wrist strap. Courtesy 3M

ESD awareness and the need for wearing wrist straps by all persons handling sensitive components at all times.

A wrist strap consists of a flexible, expandable cuff or band that encircles the wrist and a detachable conductive cord for grounding the wrist band to ground. The straight or coiled cord includes a protective series resistor. Figure 7-2 shows four different styles of wrist bands made to provide a snug fit necessary for positive skin contact.

The style at left is made from conductive metallized cotton fibers and it is secured to the wrist with a nylon hook and loop fastener and a D ring. The style at the right is made from washable polyester fabric interwoven with silver-coated nylon thread, and it has a hinged buckle for adjustment The style at the lower right is made of the same material, but it is elastic and available in three sizes: small, medium and large. The style at the top is made with adjustable stainless-steel links and hooks like a stretchable watch band. The links are removable, so the band can be fitted comfortably to all wrist sizes.

All of these wrist bands have back plates with fasteners for attaching the ground cords. They all have interior resistances less than 10^3 ohms/square. The outer surfaces of all these wrist straps are insulated to prevent shock if the wearer accidentally contacts any live electrical circuits at the work station.

The outside surfaces of the metal stretch bands are coated with an insulating plastic. These straps are more durable than the fabric straps and they can be chemically cleaned. Metal stretch bands are recommended for use in Class 100 ESD-protected clean rooms.

The grounding cords include 1/4 watt, 1 megohm (+10 percent)

115

Personal Protection against ESD

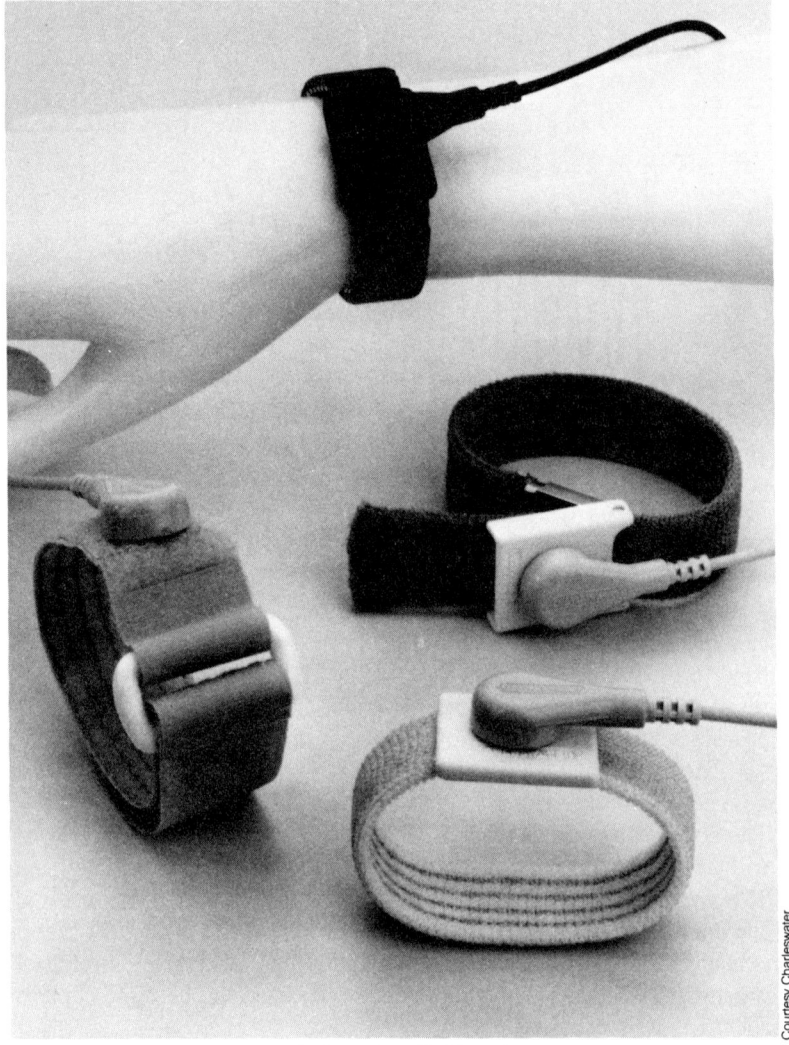

Fig. 7-2. Variety of available wrist straps.

resistors molded into one end of the cord at the wrist terminal to prevents the wrist straps from sources of electrical shock. The most frequently used resistor value is 1 megohm because it can limit the current through the wearer to an easily tolerated 1 milliampere. A resistance value of only 250,000 ohms would protect the wearer from shocks up to 1,250 volts, by limiting the current to 5 milliamperes. However, lower values of resistance can be used if the necessary precautions are taken to be sure the wearer will not contact a voltage source at the workstation that is capable of causing more than a 5-milliampere current to flow through that person.

Grounded Wrist Straps

When handling extremely sensitive devices, it is desirable to minimize the series resistance to ground because this reduces the residual voltages built up on people performing static-generating motions. However, lower series resistance also exposes persons to higher shock levels. For example, where gallium arsenide (GaAs) diodes, transistors, and integrated circuits are being handled, the value of the series resistor can be only 100,000 ohms. This sensitivity make precautions that much more important.

The series resistor in the ground lead is usually molded in a snap-on housing at the cuff end of the lead; if it were at the other end, the lead could short to ground ahead of the resistor, thereby shunting the resistance. Strain relief at the wrist and ground connections (see FIG. 7-2) ensures longer life for the wrist strap.

Properly made and maintained wrist straps have no exposed metal or conductive plastic parts. This additional safety precaution is not significant for most component inspection, handling, and assembly work because voltage sources are not present. However, covered wrist straps are necessary in test areas when the wearer is working close to, or has his hands inside, energized equipment while troubleshooting or making repairs (see FIG. 7-3).

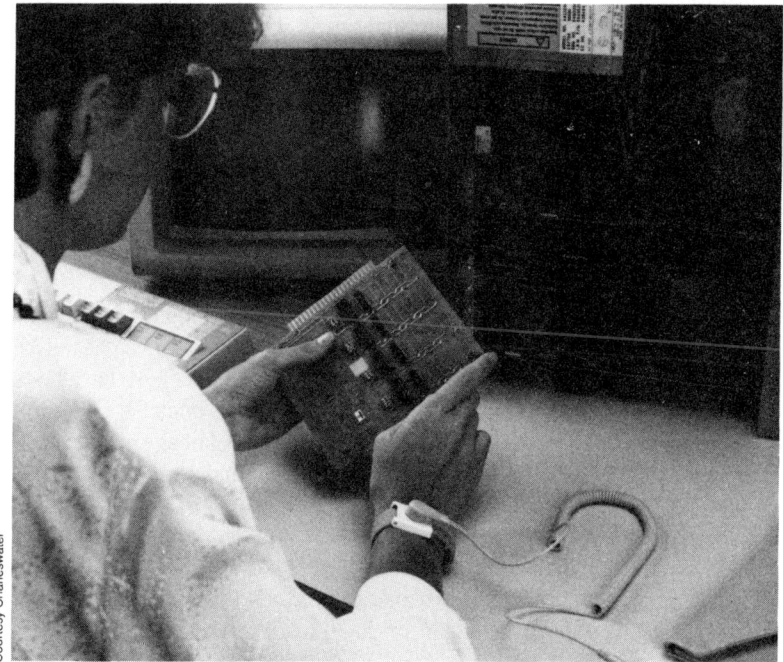

Fig. 7-3. Wrist strap used in positioning an ESD-sensitive component.

Personal Protection against ESD

A suitable ground cord is made of seven-strand cadium-copper-tinted tinsel with polyurethane insulation. It is lightweight and typically about 0.1 inch thick. To equalize the potentials between the operator and the work surface, the wrist strap cord should be connected to the benchtop at a common terminal with the ground cable. This connection can be an alligator clip, a banana plug, a snap fastener, or a heavy-duty insulated bulldog clip.

All of these devices will release the ground cord easily, without injuring the wearer, if the person should leave the workstation in a hurry—during an emergency or when simply forgetting to remove the wrist band. For the same reason, the ground cord should have an easy release connection at the contact point on the strap. However, neither release should be so tenuous that the wearer would be unaware of a break in the connection.

In selecting wrist straps, the wearer's acceptance of the fit, feel, and style is important. If the strap is uncomfortable or dirty from previous use, an employee will find excuses for not wearing it. The effectiveness of all ESD protection will be compromised if employees do not wear their straps at all times when in a controlled area.

Because wrist straps are worn close to the skin, many people are uncomfortable about exchanging them with others and prefer personally assigned units. These straps are often collected by a supervisor for central storage at the end of a work shift, tested for continuity, then returned at the start of the next shift. For situations in which it is inconvenient to use a conventional wrist ground strap, disposable grounding wrist straps (see FIG. 7-4) are available.

Wrist straps should be worn on the left wrists of right-handed persons and on the right wrists of left-handed persons—as is typical for the wearing of a watch. After a wrist strap is put on, the wearer should touch the grounded benchtop first, before handling any ESD-sensitive products. Although grounded, the person still should avoid touching any electrical leads or contacts.

Although a strong case has been made for the continuous wearing of a wrist strap by anyone working in an ESD-controlled area, it might not be so obvious why all visitors should also wear wrist straps. There have been many reports of damage and destruction to electronics because ESD-protection rules were not enforced with visitors. The damage was caused when the visitors failed to heed warnings about touching ESD-sensitive products.

Because wrist straps are subject to wearout from frequent use, they should be checked daily for continuity before each work shift. The straps must have no temporary or permanent open connections. They should also be rechecked during the shift or day's work. Wrist-strap testers are

Grounded Wrist Straps

Fig. 7-4. Disposable grounding strap provides ESD protection for a one-time user.

Personal Protection against ESD

available for checking the strap's resistance by itself or with the combined resistance of the strap and wearer. These testers are discussed in chapter 8.

FOOT GROUNDING STRAPS

A restraint on the wrist might inhibit the movement of personnel in circuit board testing areas because the technicians must stand up and move frequently. If a wrist strap cannot be worn conveniently, foot or shoe grounding straps should be used. Properly designed and made heel- and sole-to-ground straps should dissipate human voltages to zero in less than 0.1 seconds in all levels of humidity.

Conventional street shoes have soles of leather, composite materials, or rubber. Because of their much lower electrical resistance, leather shoes are recommended for use in ESD-protected areas; shoes with rubber or composite soles should never be worn in these areas. Simple shoe grounding straps can be used when personnel are wearing shoes with insulative soles. The resistance of the strap is sufficient to protect a person from dangerous electrical shocks while allowing static charge dissipation. The strap is tightened around the sole of the shoe, and the ends are inserted between the foot and shoe.

To ensure that there is proper ground contact at all times, even when wrist ground straps are used, foot grounding straps might be required. These permit proper grounding, regardless of the shoe sole's composition. An electrically conductive shoe grounding strap is shown in FIG. 7-5. This strap is especially important when the flooring has a surface resistivity of less than 10^9 ohms/square. Foot grounding assemblies are qualified to the provisions of MIL-B-81705C, method 4046 DOD-STD-1686, DOD-HDBK-263, and NFPA 56A.

There are two general types of foot grounding assemblies: the heel-ground strap and the toe-ground strap. Both are designed to establish and maintain a continuously conductive path between mobile personnel and conductive flooring. These straps are typically made of a conductive polymer with antislip properties to ensure instantaneous static dissipation at all humidity levels.

Many heel grounders are basically conductive cups or slings with elastic straps to fit all types of men's and women's low, flat-heeled shoes. A replaceable conductive fabric tab is placed inside the shoe to provide the skin-to-strap grounding path. Alternate versions of this design include a 1-megohm resistor. Another form of the heel-ground strap has a replaceable conductive calf strap, or garter, with a Velcro closure to provide skin contact. A 1-megohm resistor with a ground cord connects the garter to the heel for safe grounding of personnel. One heel ground strap should be attached to each shoe.

ESD-Protective Clothing

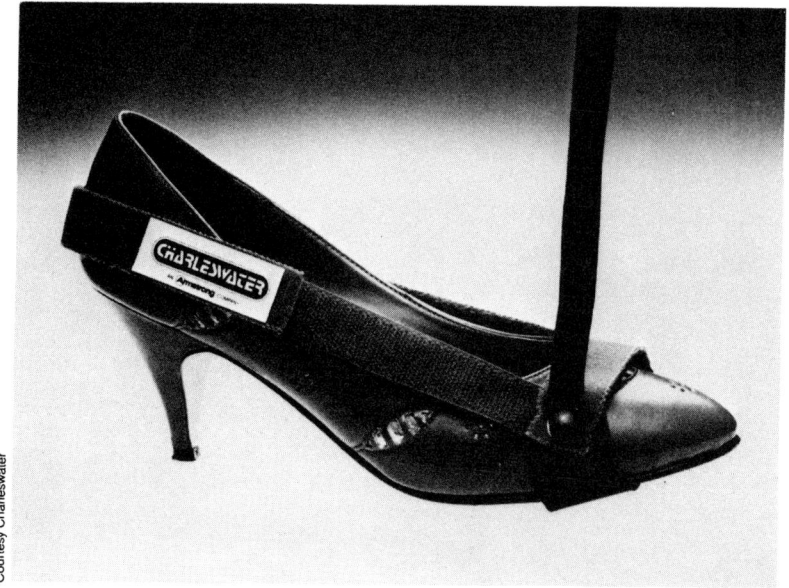

Fig. 7-5. Conductive toe strap fits high-heeled shoe.

The toe ground is a foot-grounding system made for pointed or high-heeled shoes. Also it typically is made of a conductive polymer held in place with stretch straps. A replaceable conductive tab, inserted in the shoe, provides the grounding path. Some versions of this system use 1-megohm resistors. As with the heel ground strap, one toe-ground strap should be attached to each shoe.

ESD-PROTECTIVE CLOTHING

Because the human body is a conductor, it can be effectively grounded. Clothing made from synthetic fabrics, however, will generate and retain static charges that can reach 30 kV. Unfortunately, these static charges cannot be drained or dissipated effectively by conventional grounding methods.

In any work area where ESD-sensitive components and circuit boards are handled, outer clothing worn by personnel will influence ESD control. Examples of special antistatic garments made for use in ESD-controlled areas are shown in FIG. 7-6. If, in addition to being an ESD-protected area, the work area is a clean room, clothing would also include coveralls, head covers, gloves, and finger cots.

Most certified clean room installations have adjacent clothing change rooms and special air-lock entrance and exit ways, making it relatively easy to enforce rules about wearing approved clothing. By contrast, some ESD-controlled areas might not be enclosed with restrictive entrances

Personal Protection against ESD

Fig. 7-6. Antistatic smock and shop coat.

and exits. These open areas make it more difficult for supervisors to enforce clothing codes.

In any ESD-controlled area where the standards permit street clothing to be worn without being covered with a static-dissipative shop coat or smock, it is advisable to teach people about acceptable outer clothing. Unacceptable clothing includes nylon or polyester shirts or blouses. Nylon and Dacron garments have been widely used in clean rooms because they are essentially lint free. However, they are unsuitable in ESD-protected areas because large static charges can build up on them. These garments can be used if treated with an antistatic during laundering, but these garments are far from ideal (see TABLE 7-1).

ESD-Protective Clothing

Table 7-1. Antistatic Treatment Techniques for Woven and Nonwoven Fabrics.

Technique	Advantages	Disadvantages
Disposable garments	1) Good to 20% R.H. 2) Very low initial cost	1) Possible particulates 2) Low abrasion resistance 3) Present materials cannot be reused in low R.H. environment
Dacron* with woven nylon conductive fiber	Permanent	1) Higher initial cost 2) Acid erosion a minor problem
Dacron with topical	Low cost	Fails to meet standard specifications at lower humidities
Dacron with experimental topical treatments	Meets specifications at low humidity	Currently expensive
Spray	Not developed as a control technique	Uneven distribution

In all ESD-protected work environments, employees should prevent ESD-sensitive components and circuit boards from touching their clothing. In general, short-sleeved shirts or blouses are preferred. If long-sleeved clothing is worn, the sleeves should be rolled up or covered with antistatic sleeve protectors called *gauntlets*. Made with conductive fabric or plastic sleeves that reach from wrist to elbow, gauntlets have elasticized cuffs that fit snugly around the bare wrist. With gauntlets, shirts and blouses made of synthetic fabrics can also be worn by employees.

Conventional street clothing can be converted to antistatic garments by treating cotton or synthetic fabrics with an antistatic chemical agent in the final rinse during laundering (see "Topical Antistats" in chapter 6). However, each time these clothes are washed, the antistatic agent must be reapplied.

Another common type of antistatic garment is made from a fabric of 65% polyester, 34% cotton, and 1% stainless steel fiber. The steel fiber is an integral part of the fabric. The fabric can be worn, handled, and laundered like ordinary clothing. In critical cleanroom applications, however, neither chemically treated garments nor those made from fabric, including stainless steel fibers, are acceptable. Cotton garments, for example, shed too much lint, even when they are blended with synthetic fabrics.

ESD-Protective Clothing

The steel fiber eventually breaks into fine particles after prolonged use and repeated laundering.

Protective clothing should be checked frequently with an electrostatic field meter (see chapter 8), especially after cleaning, to monitor for potentially damaging static voltages. High readings will be found at the edges of the material—on creases, cuffs, and hems. Even higher readings will be found on garments made from multiple layers of synthetic materials.

GLOVES AND FINGER COTS

Latex gloves or finger cots are used in clean rooms to keep body moisture, oils, and contaminants off ESD-sensitive objects; however, both can cause ESD problems. Powder- and static-free gloves have been developed. These gloves have minute gaps between the glove surface and the hand to prevent perspiration and permit vapor dispersal.

Finger cots used in the ESD-protected work should be made from ESD-protective materials. Finger cots should be used only in an environment where ionized air is circulated.

FIELD SERVICE KIT

A field service kit typically consists of a portable grounding workstation made from flexible, static-dissipative mat material, a common-point ground cord with a removable clip, and an adjustable wrist strap with retractable cord. Figure 7-7 shows a field service kit in use. The static-dissipative mat safely drains static charges from people and conductive objects to ground on contact with the surface. The mat might contain storage pockets; still, it is designed to fold up easily for storage in a static-shielding bag. The common-point ground cord is typically about 15 feet long; the wrist strap cord is about 6 feet long.

The surface resistivity of a mat suitable for this application is typically less than 10^9 ohms/square. Static decay is typically less than 50 milliseconds, and surface-to-ground resistivity is typically less than 5×10^8 ohms. Volume resistivity is less than 10^8 ohm-cm.

Field Service Kits

Fig. 7-7. Field service kit includes a 19- × -22-inch conductive vinyl work surface.

8
Electrostatic Test Equipment

Specialized test equipment is needed to define electrostatic problems, evaluate solutions, and monitor ESD-related situations continuously. Instruments are available to measure surface resistivity, resistance to ground, electrostatic fields, ionization effectiveness, ion balance, tribocharging, and static decay. Other instruments are used to monitor the humidity and temperature in ESD-controlled rooms.

This chapter covers the instruments associated with ESD control. These instruments are generally classified according to their ability to survey existing conditions, evaluate those conditions, or monitor them continuously. Most instruments described in this chapter can be used effectively by nontechnical personnel who have been trained in their operation.

Surface resistivity, resistance to ground, and electrostatic field measurements are checked on packaging materials, work surfaces, floors, clothing, equipment, and raw materials. Ionization effectiveness and ion balances are measured in rooms and work areas. Tribocharging is measured on packaging materials and raw materials, and static decay is measured on packaging materials, clothing, work surfaces, and floors.

The general classes of test equipment described in this chapter are: 1) electrostatic fieldmeters and voltmeters, 2) electrostatic monitors, 3) surface/volume resistivity probes and meters, 4) humidity test chambers, 5) static decay meters, 6) shielded bag test systems, and 7) ground strap testers.

PORTABLE SURVEY AND AUDIT INSTRUMENTS

One instrument manufacturer has introduced portable, battery-powered instrument kits for measuring resistance, resistivity, static charge, temperature, and humidity—all key variables in the maintenance of an ESD-controlled facility that must be routinely audited. For convenience, all these instruments can be carried in a single case (see FIG. 8-1). However, it might be necessary to use a more precise fieldmeter than the one shown if the devices being handled are generally sensitive to less than 1,000 volts ESD.

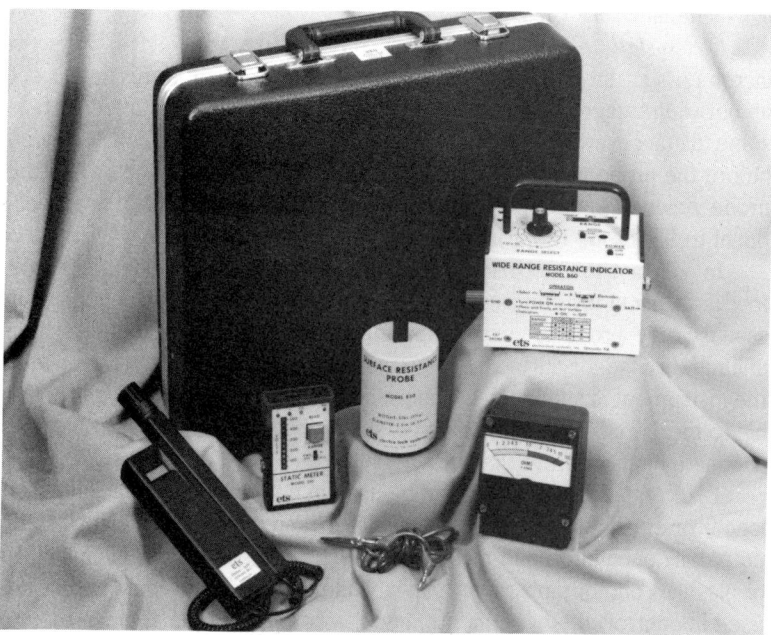

Fig. 8-1. Kit of portable instruments for measuring resistance, resistivity, static charge, temperature, and humidity.

THEORY OF ELECTROSTATIC FIELDMETERS AND VOLTMETERS

Electrostatic voltmeters and fieldmeters measure the magnitude and polarity of electrostatic charges existing on materials, objects, and people. In addition, these meters can measure the approximate magnitude of electrostatic charges generated by personnel movements and triboelectric charges generated by friction between two materials. However, these

Electrostatic Test Equipment

meters cannot detect rapid transients or pulses with fast rise times and short pulse widths.

Both electrostatic voltmeters and fieldmeters are used for electrostatic measurements. Although they have many basic similarities, they also have some differences that should be noted.

Electrostatic voltmeters provide noncontacting measurement for the electrostatic surface potential, or *surface voltage*, on insulators, semiconductors, or conductors. Their measurement accuracy is independent of probe-to-surface separation. These meters can have high accuracy—0.1 percent or better—and the ability to resolve small spots—approximately 0.1 inch diameter or less.

Electrostatic fieldmeters measure the electrostatic field in volts-cm increments at the ground probe. For most measurements, this field is set up between the grounded bottom plate of the probe and a charged surface some distance away. The free space field changes from what existed before the probe was inserted, except in certain circumstances when the probe views the field through or at the ground plane. The field is proportional to the probe-to-surface separation.

In contrast to electrostatic voltmeters, electrostatic fieldmeters have moderate accuracy. Surface voltage errors are typically 2 to 5 percent in cases where probe-to-surface spacing is established. Fieldmeters have unlimited range for surface voltage measurements. Because the range of the instrument is a function of the probe-to-surface separation, it is possible to measure hundreds of kilovolts simply by positioning the probe at an adequate distance from the surface under test (SUT).

Electrostatic fieldmeters can be used in both production facilities and laboratories. These instruments can detect and measure the electrostatic potentials encountered in high-speed machines used for the manufacture of plastics, photographic film, paper, and textiles. These fieldmeters are used in controlled environments, such as computer rooms, clean rooms, and hospitals.

When selecting an electrostatic fieldmeter consider the following characteristics:

- Suitability of the instrument for the end-use application
- Analog or digital display
- Readability of the display
- Inherent measurement accuracy (at least ± 10 percent)
- Repeatability
- Sensitivity in terms of minimum voltage level that can be accurately measured
- Range of measurable voltages
- Electrometer or chopper-stabilized sensor
- Hand-held or bench-type operation
- Simplicity of operation

Theory of Electrostatic Fieldmeters and Voltmeters

Types of Fieldmeters

The three basic variations of electrostatic fieldmeters are: the electrometer-type, the radioactive-type, and the ac carrier-type.

The Electrometer. The *electrometer*, or pocket-size electrostatic locator fieldmeter, is basically a capacitively coupled dc amplifier with a shunt capacitor for calibration. Its operation is based on a parallel-plate capacitor. The meter probe becomes one plate of the parallel-plate capacitor. The object, whose field is being measured, is the other plate. The voltage is induced from the field being measured to the probe of the meter. This charge, which is proportional to the field, is measured by the dc amplifier.

This type of fieldmeter is inexpensive, simple, and small, and measures rapidly. Its disadvantages include an inability to monitor an area over an extended period of time, the need to zero it periodically, and a drift in its dc-coupled circuitry. The amplifier is also subject to imbalance if the meter is used in an ionized air environment.

The Nuclear Fieldmeter. The second type of fieldmeter uses a radioactive or nuclear sensor. The electrostatic field induces a charge in a grid, located at the tip of the meter, that is proportional to the charge being measured. An ionizing source, such as tritium foil, causes the air space between the grid and a high-impedance amplifier to become electrically conductive. As the air is ionized, a small current that is proportional to the electrostatic field flows from the grid to the amplifier. The current is then measured to determine the magnitude and polarity of the electric field. The spacing between the object and the tip of the meter determines full-scale sensitivity.

These fieldmeters are very simple and their dc amplifiers are stable. However, many companies have stopped purchasing them, as well as nuclear ionizers, because their employees have expressed fears of any instruments containing radioactive materials. This heightened awareness of the dangers of radioactivity has been attributed to the well-publicized nuclear reactor accidents in the United States and the Soviet Union.

Despite the fact that no health problems have been attributed to proximity to or from the use of nuclear fieldmeters, they have been replaced by other types of fieldmeters to avoid employee complaints. The instruments also have other drawbacks, including possible reading errors caused by dust accumulating on the radioactive source and the need for frequent replacement of the source material because it has a relatively short half-life.

The ac-carrier Fieldmeter. The *ac-carrier fieldmeter*, also known as the *chopper-stabilized fieldmeter*, is the version most commonly used for making high-quality electrostatic field measurement. The ac signal is produced by modulating a capacitive pickup in an electric field with a rotary chopper or a vibrating capacitor.

Electrostatic Test Equipment

The rotary chopper version contains a rotor that "chops" the electric field periodically. The sensitive electrode might be the stator behind the rotor or a separate capacitor plate behind both the rotor and stator (see FIG. 8-2). The aperture moves between the charged surface (one plate of capacitor C1) and the fixed sensor (the second plate of the capacitor C1).

As an impellor or shutter rotates, blades pass in front of the sensor. During this cycle, the charge is alternately induced on the probe, then blocked or *chopped*, by the aperture through grounded contact with the case and the operator. This modulation interrupts the field and causes the sensor plate to produce an electrical output. The signal's amplitude represents the field intensity, and the signal's phase indicates the field polarity.

Although the design of the chopper-type fieldmeter is simple, this instrument is larger than the nuclear and electrostatic types because it uses a shutter drive motor. Also, the speed of response for this instrument is limited by the motor speed.

A second version of this fieldmeter employs a vibrating capacitor. The sensitive capacitor electrode is vibrated perpendicularly, with respect to the electric field on the surface, to be measured by a modulator driver. An aperture in the bottom plate of the probe permits the entry of the field. The amplitude of the ac signal created is proportional to the modulation amplitude. Therefore, the modulation amplitude must remain stable to obtain accurate readings. Null-seeking feedback techniques are used widely to compensate for modulation amplitude changes.

These instruments can be made rather small, with probe sizes ranging from 1 to 2 cubic inches. Extremely reliable, these meters draw little power. The speed of response is typically 0.5 second; they are used widely for long-term monitoring.

Electrostatic Fieldmeters

The Monroe Model 171 electrostatic fieldmeter (FIG. 8-3) measures electrostatic field strength or intensity—*potential gradient*—up to ± 10 kV/cm. It can also be used to determine surface voltage by using the probe-to-surface separation as a calibration factor. This instrument is expandable from 2 to 16 channels. Each channel has two alarms brought out separately: one to indicate a positive field beyond the set limit, and the other a negative field beyond its limit. These lines activate alarm relays. A meter is provided only for operator convenience and local viewing because the Model 171 is designed to feed data to a central data collecting/monitoring computer through analog outputs.

The Model 171 is principally used to monitor electrostatic charge accumulation at up to 16 points simultaneously. As the charge increases on the surface of a material, the electrostatic field in the vicinity also

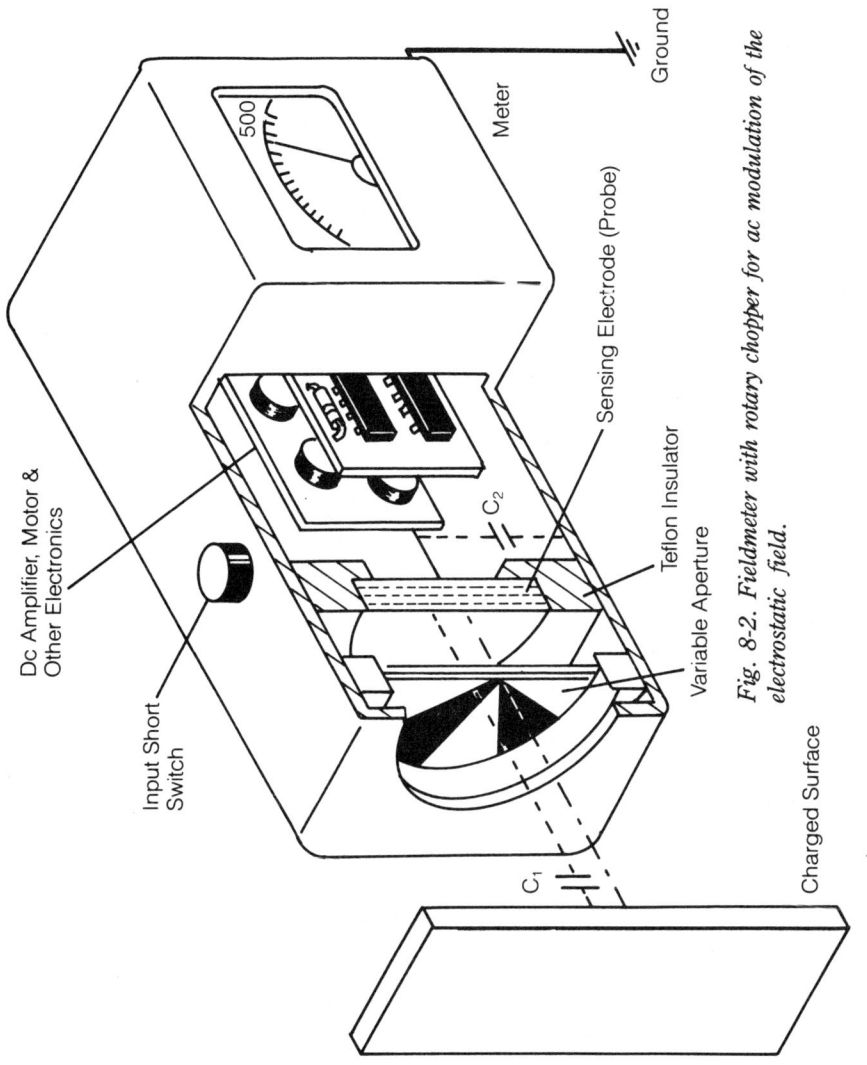

Fig. 8-2. Fieldmeter with rotary chopper for ac modulation of the electrostatic field.

Electrostatic Test Equipment

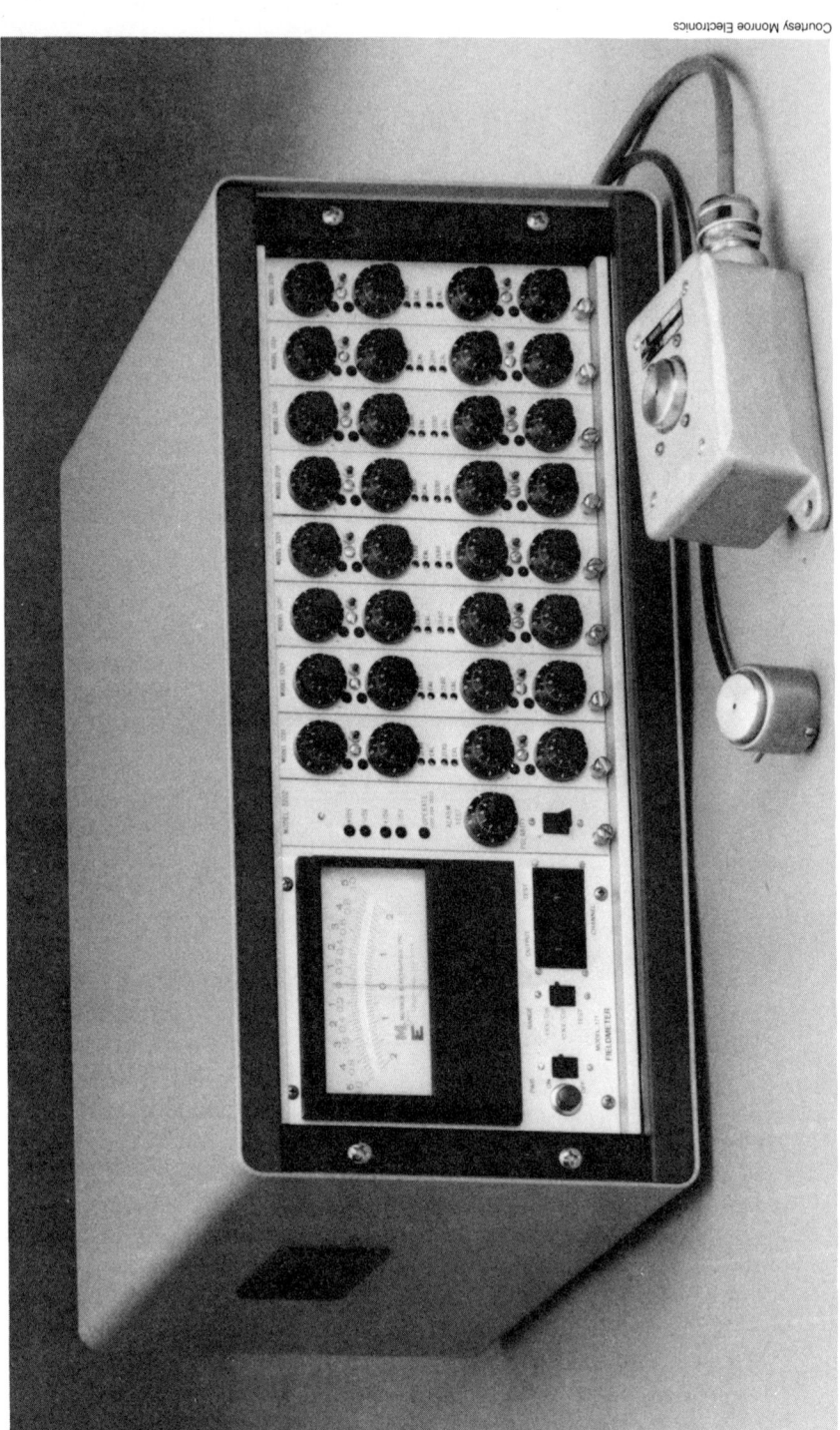

Courtesy Monroe Electronics

Fig. 8.3 Electrostatic fieldmeter

increases proportionally. Thus, the Model 171 provides an output signal directly proportional to the surface charge accumulation, while making no physical contact with the material being monitored.

The probes are intended for use anywhere charge buildup is monitored. The probes can be separated from the mainframe computer by up to 1,000 feet, but they must be close to the surface being measured.

A simplified block diagram of the Model 171 fieldmeter is shown in FIG. 8-4. A vibrating sensitive electrode—capacitor plate—*looks* at the SUT to be measured through an aperture at the base of the probe assembly. The modulated ac signal induced on this electrode is proportional to the differential voltage between the SUT being measured and the bottom plate of the probe assembly. Its phase is determined by the polarity of the dc voltage on the SUT.

This ac signal, conditioned by the preamplifier, filter, and signal amplifier within the probe, is coupled into a phase-sensitive demodulator. The signal from the demodulator is then fed to an integrator circuit. A voltage from the integrator is fed back to the probe assembly to null the field. When the field is nulled, the integrator output stabilizes. When this occurs, the voltage output of the integrator is directly proportional to the field intensity at the probe. This signal is then directly measured by a voltmeter to provide a direct readout. The mechanical modulator-driver is driven by an oscillator whose frequency is determined by the phase-sense demodulator.

Hand-held Electrostatic Fieldmeters and Voltmeters

Hand-held electrostatic fieldmeter units can be packaged in rectangular pocket-type cases with a moving-coil meter, (FIG. 8-5), an analog-type LED bar indicator (FIG. 8-6), or LCD displays. Some even have pistol-grip cases for easier holding and pointing. These instruments are also called static meters, electrostatic locators, static locators, field scanners, electrostatic detectors, electrostatic meters, and static electricity detector/monitors.

The accuracy of these meters is less than that of the bench-type meters. However, because they are small and light enough to be carried by hand easily, these meters are useful for making routine repetitive measurements and determining the presence, polarity, and rough magnitude of electrostatic charges on various surfaces. These fieldmeters can spot problem areas, which can be later analyzed by more precise instruments. As discussed earlier, these instruments, unlike conventional voltmeters, are noncontacting. (An inadvertent contact between the meter and the surface could discharge the surface to be measured.)

Hand-held meters are usually powered with disposable 9-volt alkaline batteries, although some use rechargeable NiCad cells. Some of these

Electrostatic Test Equipment

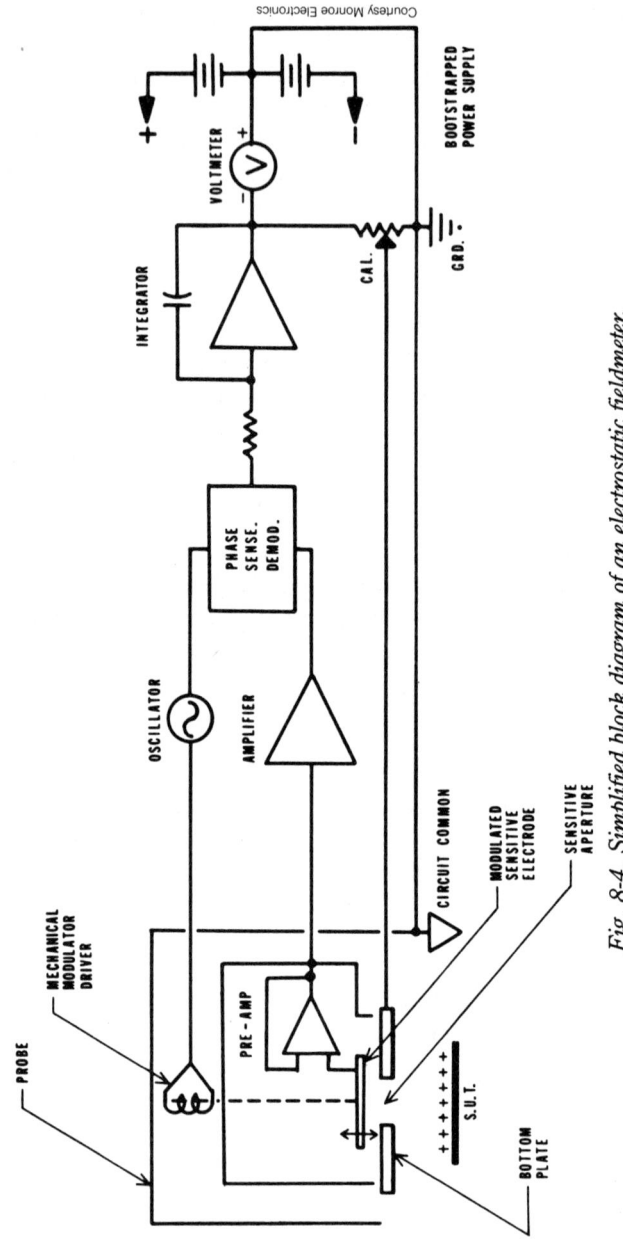

Fig. 8-4. Simplified block diagram of an electrostatic fieldmeter.

Theory of Electrostatic Fieldmeters and Voltmeters

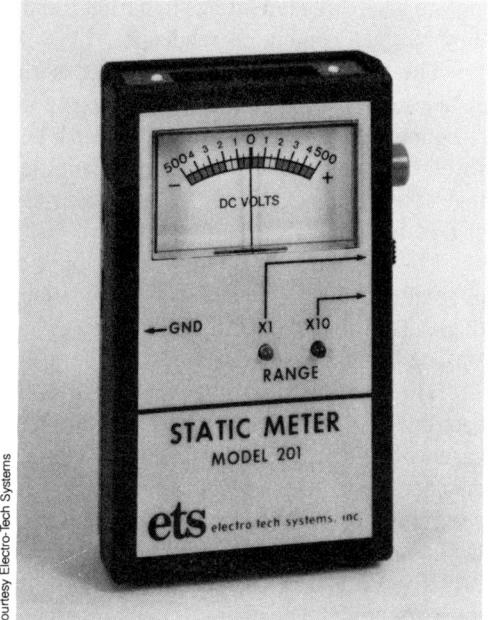

Fig. 8-5. Static meter with analog moving-coil meter.

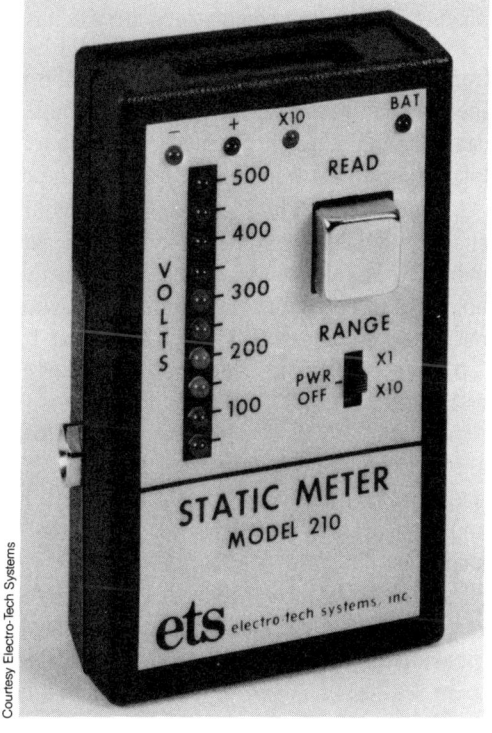

Fig. 8-6. Electrostatic fieldmeter with LED bar graph.

Electrostatic Test Equipment

meters have threaded fittings to permit them to be mounted on a tripod so they can take continuous readings.

The sensor end of the fieldmeter is aimed—from a predetermined distance, depending on the voltage range potential involved—at the object to be measured. The operator activates a button or switch, and the measured value appears on the meter, readout, or indicator. The value usually registers in units of volts or kilovolts per meter and can either be positive or negative.

Most portable analog meters cannot be read directly across the complete measuring range, and require reference to multipliers or switch settings. The accuracy of these meters depends on the distance to the surface being measured.

The normal distance between the detector and the target varies with each meter, but typically it is 0.5 to 24 inches. The distance might also vary with different scales on the same meter. Failure to keep the specified distance can result in reading errors. For this reason, movable distance-setting guides might be used to ensure that the specified distance is maintained between the meter and the target.

It is necessary to refer all meters to the zero field before making measurements. The operator must be grounded with a wrist strap before using the instrument to maintain a zero potential. Otherwise, simple motions such as scuffing shoes on the floor might introduce errors in the reading. (Wrist grounding straps are discussed in chapter 7).

Any surface to be measured should be positioned in free air, away from a ground plane, since a ground plane near a surface will terminate the field and cause erroneous readings on the field meter. A triboelectrically charged surface produces different readings in free air than those taken next to a ground plane, although the charge is the same.

Model 265. The Model 265 static meter from Monroe Electronics is chopper-stabilized for making survey measurements in ionized environments. It is a pocket-sized fieldmeter for quick measurements that does not need to be continually rezeroed. A switch-selectable dual range permits measurements of 0 to 1,000 volts at 1 inch from a surface and 0 to 3,000 volts at 4 inches. In the 10X range, it can measure 0 to 10,000 volts at 1 inch and 0 to 30,000 volts at 4 inches.

Model 255. For evaluating electrostatic fields, Monroe offers another hand-held fieldmeter: the Model 255. Also chopper-stabilized for use in ionized environments, it features a range of $\pm 20,000$ volts/inch, and measures kV/inch with respect to the grounded case. Accuracy is said to be 5 percent.

Model 201. The Model 201 static meter from Electro-Tech Systems is a noncontacting hand-held meter designed to locate and accurately measure the magnitude and polarity of electrostatic fields. This instrument (FIG. 8-5) has two voltage ranges that can be set manually. It is cali-

Theory of Electrostatic Fieldmeters and Voltmeters

brated to provide full-scale readings of ±500 volts or ±5,000 volts when the meter is held 2 inches from a charged surface. Meter sensitivity can be increased by moving it closer to the charged surface, allowing full-scale measurements of less than ±500 volts to be made. By moving the meter farther from the charged surface, electrostatic fields in excess of ±5,000 voltscan be measured.

The 1X or 10X range is selected with a two-position switch. When the operating button is depressed, green or red LED lamps show the selected range. The manufacturer claims that the easily read analog moving-coil meter avoids the repetitive sampling of readings that occurs in similar units with digital displays. Powered by a standard disposable 9-volt alkaline battery, the meter has a low-battery circuit that turns off the LED range indicators when the voltage has dropped below the preset minimum. The Model 201 is calibrated with a 12- x -12-inch metal plate held 2 inches from the meter.

Model 210. The Electro-Tech Model 210 static meter (FIG. 8-6) employs a ten-LED analog bar graph display. This meter indicates the magnitude and polarity of electrostatic fields by selectively illuminating LEDs. Other LEDs indicate the selected ranges and low-battery condition.

Electrostatic Voltmeters

The Monroe Electronics Model 244 electrostatic voltmeter (FIG. 8-7) permits accurate measurements of electrostatic or other high-impedance sources without physical contact. As shown in the simplified block diagram, FIG. 8-8, the Model 244 employs an electrostatic chopper for low drift, and negative feedback for accuracy. It has a range of 0 to ±3,000 volts dc.

The surface resolution of this instrument depends on probe aperture size and probe-to-surface spacing. To obtain the best performance, probe-to-surface spacing should be kept as close as physically reasonable. The typical spacing range is from 0.005 inch for unknown voltages below 500 volts to over 0.125 inch for unknown voltages up to 5,000 volts. As probe-to-surface spacing increases, instrument accuracy, speed of response, and surface area decreases, and noise and drift increases.

The electrostatic electrode "looks" at the surface being measured through a small hole at the base of the probe assembly. The chopped ac signal induced on this electrode is proportional to the differential voltage between the surface being measured and the probe assembly. Its phase is determined by the dc polarity.

The reference voltage and the mechanically modulated signal, conditioned by the high input impedance preamplifier and signal amplifier, are fed to a phase-sensitive detector. The polarity and amplitude of the dc

137

Electrostatic Test Equipment

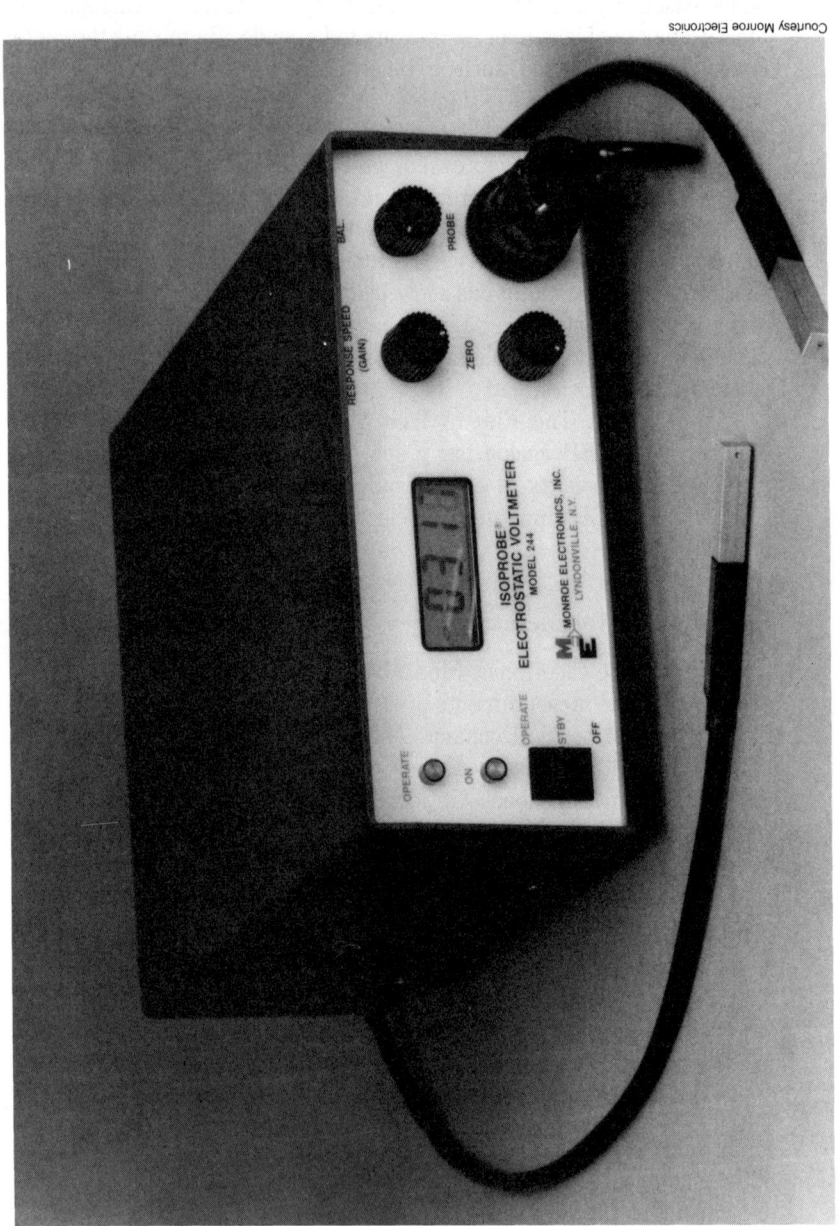

Fig. 8-7. Electrostatic voltmeter.

Theory of Electrostatic Fieldmeters and Voltmeters

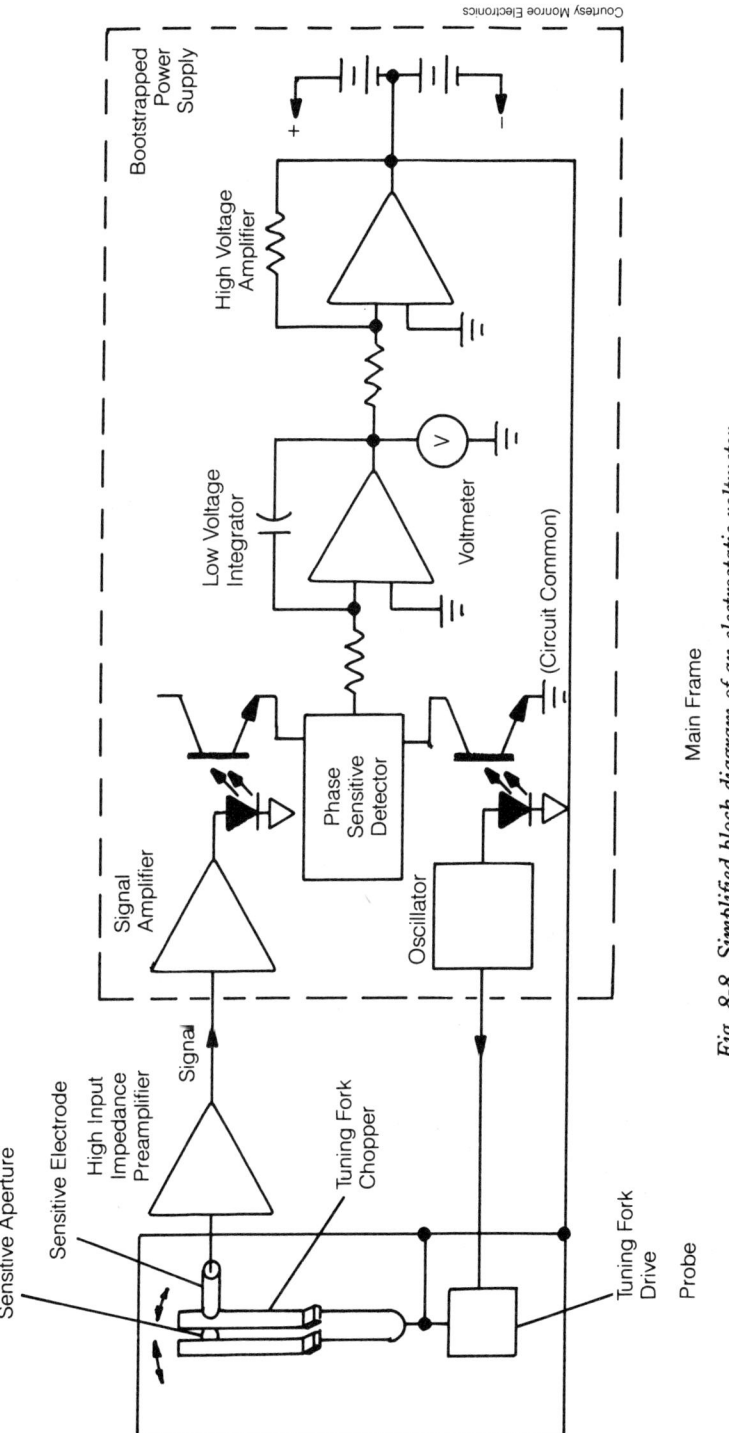

Fig. 8-8. Simplified block diagram of an electrostatic voltmeter.

Electrostatic Test Equipment

output are determined by the amplitude and phase of the electrostatically induced signal relative to the reference signal. The output of the phase-sensitive detector feeds a dc integrating amplifier. Its output polarity is inverted to that of the unknown. The output of this amplifier drives a high-voltage amplifier, which drives the probe to the same potential as that of the surface being measured.

The probe is driven to a dc voltage, typically within 0.1 percent of the potential of the unknown for a 0.04-inch probe-to-surface spacing. An accurate indication of the unknown potential can be obtained from the instrument.

The Monroe Model 244 voltmeter is a noncontacting voltage follower. The potential of its probe will attempt to follow the potential of any object within its field of view. If the meter indicates 1,500 volts, for example, its probe will be operating at 1,500 volts above ground.

Gradient adapters are available as accessories so the Model 244 can be used as an electrostatic fieldmeter. It will give a direct readout in volts/mm of separation between the SUT and the grounded plate of the adapter. The adapter consists of a 1-inch-square gold-plated metal plate with an aperture for probe *viewing*, a 10-foot flexible grounding wire with clip, and a small insulated box designed to slip over the end of the probe.

ELECTROSTATIC MONITORS

Electrostatic monitors are designed to sound an alarm that will alert personnel when the electrostatic potential in the vicinity reaches a predetermined dangerous level. These monitors permit preventive action before any ESD damage occurs. Some monitors are available as individual self-contained units; others are multistation units with several sensors connected to one central control. The sensors can operate as far as 1,000 feet from the main console. In addition to sounding an alarm, the console might automatically activate air ionizers and provide a strip-chart record of the event.

Another type of electrostatic monitor from Monroe Electronics, the Model 248 charged body detector, is shown in FIG. 8-9. This instrument detects electric fields from moving charged objects or people by using special multilayer tape sensors. It can trigger any of several alarm modes if the voltage exceeds a preset level, which is adjustable from 10 to 10,000 volts. By connecting a counter to a relay output, the user can track the number of ESD events or charged object intrusions per day. Analog output for recorders is provided by the detector; it is suitable for workstation monitoring to evaluate the effectiveness of ESD control. The flexible Model 1021 tape sensor can be looped or folded around corners and doorways.

Electrostatic Monitors

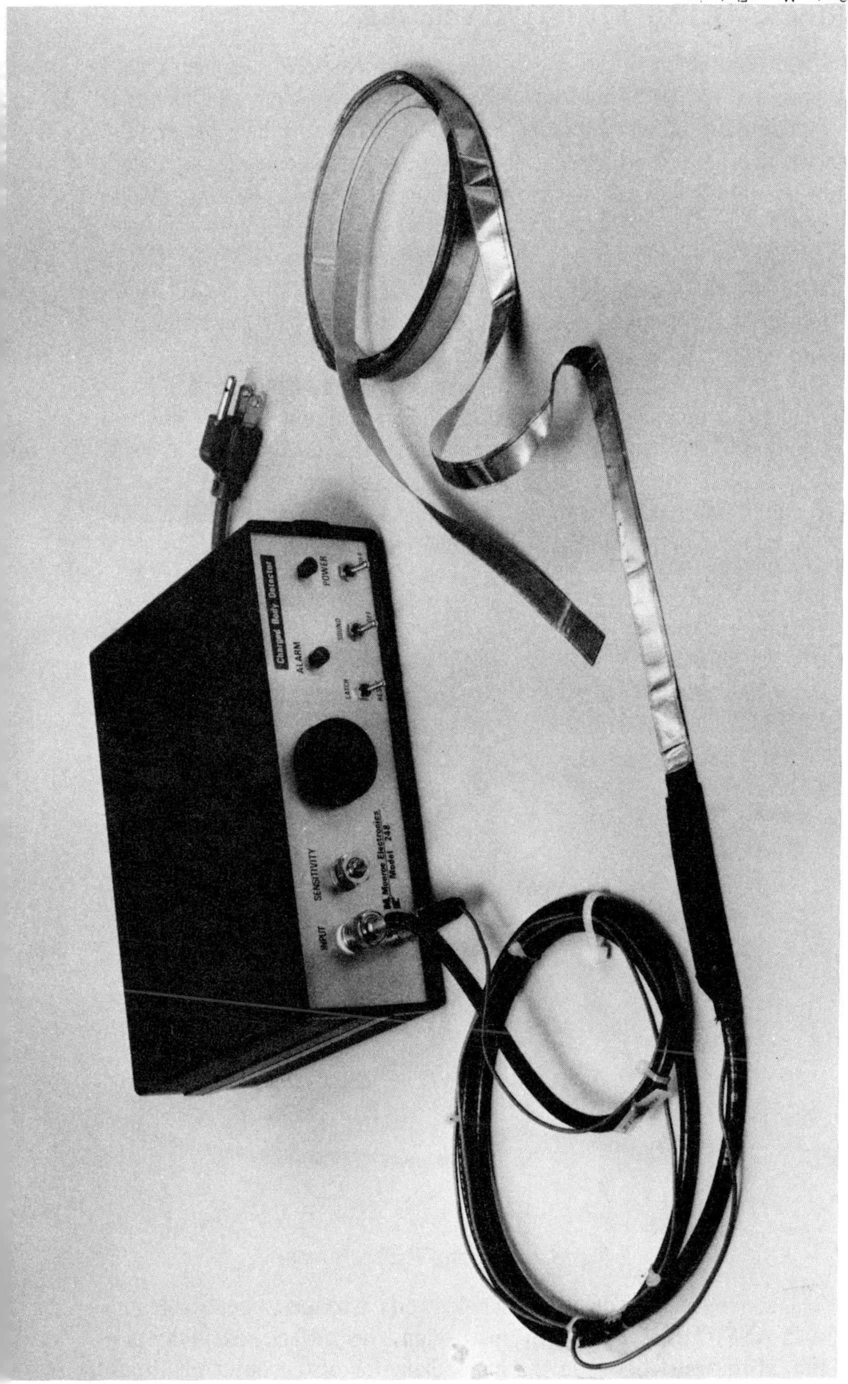

Fig. 8-9. Charged-body detector.

Electrostatic Test Equipment

SURFACE/VOLUME RESISTIVITY PROBE

Static dissipation is directly related to the resistivity of materials. This property should be measured during the development, evaluation, and qualification of all static-dissipating or static-shielding materials. Materials that are coated, chemically treated, or contain internal antistatic agents are generally only surface-conductive. However, materials filled with conductive ingredients, such as carbon, are both surface- and volume-conductive. Specifications and standards—including ASTM-D-150, ASTM-D-257, EIA-541, MIL-B-81705, EOS/ESD-STD-4, and NFPA 99—require that resistivity be measured as one of the electrostatic properties of materials.

The Model 803A surface/volume resistivity probe from Electro-Tech Systems (FIG. 8-10) measures the surface and volume resistivity of any relatively smooth, flat material with a diameter of at least $2^{1}/_{2}$ inches. This specimen size is compatible with the standard specimens used in most electrostatic material evaluations. The cylindrical probe provides a 10X multiplication factor to the surface-resistivity measurement.

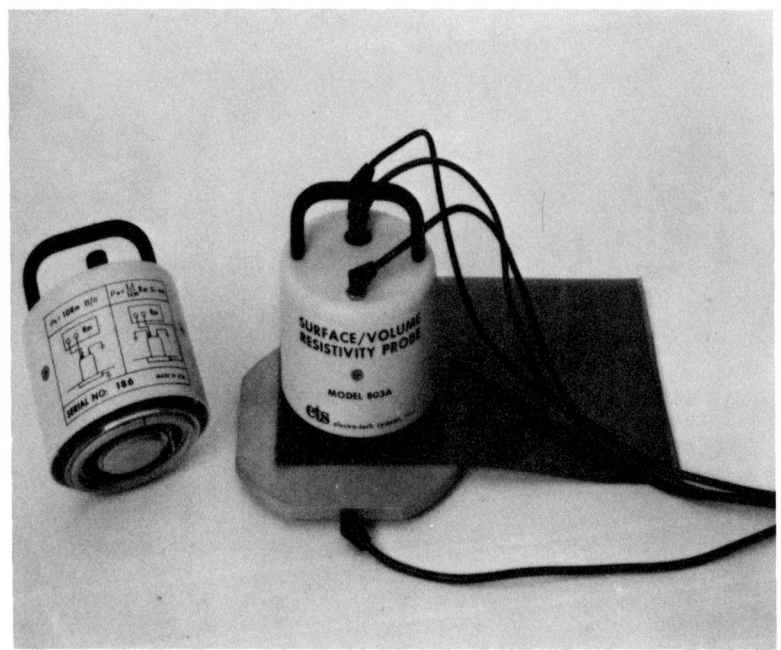

Fig. 8-10. Surface resistivity probe.

The concentric ring design of the electrode is based on applicable formulas in ANSI/ASTM-D-257. In this design, the surface resistivity is a function of the ratio between the inner diameter of the outer electrode and the outer diameter of the inner electrode. To measure surface resistivity, the test specimen is placed on a ground plane and the resistance between the two electrodes is measured. Surface resistivity, ϱ_s is then:

$$\varrho_s = 10\ R_m\ \text{ohms/square}$$

Where: R_m = ohms

Volume resistivity is a function of area, A, of the inner electrode, the thickness, t, of the test specimen and the measured resistance through the material. To measure volume resistivity, the ground plane becomes the return electrode, while the outer electrode is grounded and becomes the guard electrode. The volume resistance of the test sample is measured and the volume resistivity, ϱ_v, is calculated using the formula:

$$\varrho_v = \frac{A}{t} R_m\ \text{ohms} - \text{cm}$$

Where: $A = \text{cm}^2, t = \text{cm}$

The 803A probe is an assembly of brass concentric rings contained in a 3 1/4-inch-diameter, 4 1/2-inch-high drawn steel cup. This outer case acts as a shield when making high-resistance measurements. The internal electrodes are mounted on Teflon™ insulators, which provide high insulation resistance between electrodes under relatively high humidity conditions. The electrode contact material is silver-impregnated silicone rubber with a resistance of less than 1 ohm. The complete probe weighs 5 pounds.

Standard accessories include a 5-inch-diameter ground plane, an acrylic insulated plane, and cables for connecting the probe to a resistance meter. Any meter capable of measuring resistances over the desired range can be used with the Model 803A probe.

STATIC DECAY METER

A *static decay meter* measures the ability of a material, when grounded, to dissipate a known charge that has been placed on its surface. This meter tests in accordance with a method to evaluate antistatic-treated materials described in Federal Test Method Standard 101C, Method 4046, *Electrostatic Properties of Materials*. Static decay testing is specified in MIL-B-81705, EIA-541, and NFPA 99.

The Electro-Tech Systems Model 406C static decay meter (see FIG. 8-11) is a completely self-contained system for measuring the electrostatic dissipation or decay characteristics of materials. The meter is said to meet the applicable test requirements for static dissipation from the DOD, NFPA, and EIA.

The instrument system consists of two components: the control unit, and the Faraday test cage. This configuration enables the test cage to be

Electrostatic Test Equipment

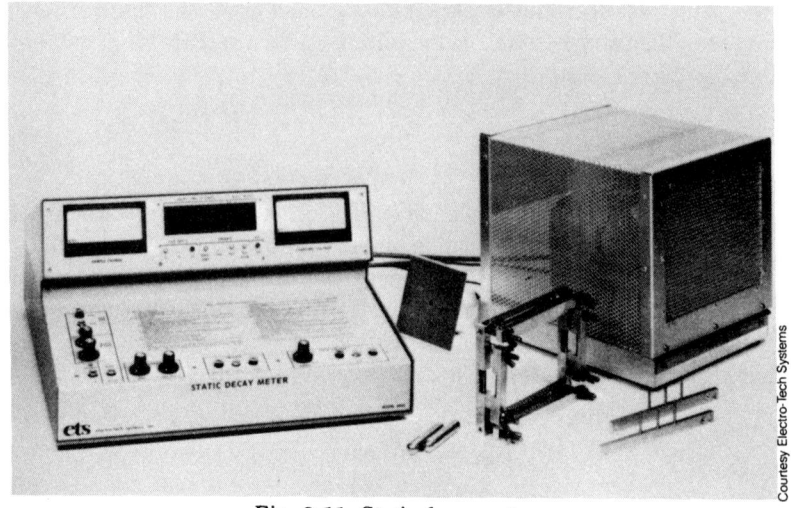

Fig. 8-11. Static decay meter.

placed within a humidity chamber for material testing under controlled environmental conditions.

A control unit (on the left in FIG. 8-11) contains an adjustable 0- to ±5,000-volt power supply and a precision electrostatic voltmeter with analog moving-coil meters for indicating charging voltages and sample charges. A four-digit LED digital readout displays decay time up to 99.99 seconds with a 0.01-second resolution. Pushbutton switches permit the selection of cutoff levels of 0, 10, and 50 percent for decay time measurement. The system can be operated either manually, with all functions controlled by the operator, or automatically for repeated, consistant testing of the sample.

The Faraday test cage shields the test sample from extraneous electrostatic fields. It contains the sample holder electrodes and the electrostatic sensor. A safety interlock is activated when the cage cover is open. A test module is supplied to verify that the instrument is operating correctly. This module contains electronic circuitry that simulates the static decay characteristics of a test sample.

Several types of standard sample holder electrodes are available for the Model 406C, permitting the user to test a wide range of materials in various sizes and shapes. For example, magnetic electrodes are used for thin, flexible film and fabric samples; clamp electrodes are used for nonflexible sheet and foam samples up to 1 inch thick; and IC tube electrodes permit the nondestructive testing of IC shipping tubes. Other standard electrodes include loose-fill electrodes for the testing of loose packing chips, as well as ring electrodes for the nondestructive testing of bottles, cups, and canisters. Custom electrodes can be designed to meet special test requirements.

HUMIDITY TEST CHAMBER

Electrostatic characteristics of materials are, in many cases, a function of ambient relative humidity. Under certain operating and storage conditions, the relative humidity (RH) can be less than 10 percent. When evaluating materials for electrostatic performance in categories such as static decay and resistivity, the testing environment must be specified and controlled. MIL-B-81705 and EIA-541 (12 $\pm$3 percent RH) and NFPA 99 (50$\pm$2 percent RH) are commonly referenced standards.

The Electro-Tech Systems Model 506 humidity test chamber (FIG. 8-12) provides a controlled humidity environment. It is designed for use wherever a controlled humidity environment is required to test, fabricate, or store critical materials or parts.

Fig. 8-12. Humidity control chamber.

The chamber is an airtight 0.25-inch-thick acrylic glove box measuring 3 × 2 × 1.5 feet. The box is accessible through a 1-foot-square door located on the left-hand side. The door can be sealed, closed, and locked. The instrumentation and test samples placed inside the chamber can be handled with a pair of neoprene rubber gloves.

The chamber is equipped with a desiccant-pump drying system. The desiccator contains the renewable drying agent, anhydrous calcium sulfate ($CaSO4$), and it is mounted outside the chamber. A small pump draws air from the chamber and forces it through the desiccator back into the chamber. The circulating system is capable of producing very low humidities within the chamber.

Electrostatic Test Equipment

A temperature/humidity indicator monitors the relative humidity within the chamber. An optional automatic humidity controller provides precise humidity level indication and automatic control.

SHIELDED BAG TEST SYSTEM

The static-shielding bag is intended to protect ESD-sensitive electronic components. The EIA-541, "Packaging Material Standards for ESD-sensitive items" specifies a test procedure for evaluating the shielding effectiveness of the static-shielding bag.

The Model 811/402 shielded bag test systems from Electro-Tech Systems (FIG. 8-13) measures the relative shielding effectiveness of metallized and other laminated films and materials in accordance with EIA-541. The system consists of a Model 811 high-voltage controller and a Model 402 shielded-bag test fixture. The measurements are performed with a two-channel oscilloscope.

The Model 811 high-voltage controller provides the necessary voltages and control signals to operate the shielded bag test fixture. The controller generates test voltages from 0 to $\pm 2,000$ volts; charging voltage rates are selectable from 50 to 2,000 volts/second in six steps. The test system can be operated either manually or automatically. A trigger output signal synchronizes the oscilloscope to the sensor output signals.

The Model 402 bag-shielded-bag test fixture consists of a bounceless charge/discharge relay, a discharge probe grounding relay, a plug-in human-body model capacitor and resistor, a discharge probe, a capacitive sensor, and attenuators for direct connection to oscilloscope amplifier inputs.

To test a bag, the differential output voltage of a capacitive sensor, placed inside a bag or in a folded sheet of static-shielding material, is measured (see FIG. 8-13). The test sample is placed on a ground plane and the discharge probe is placed on top. A human-body model ESD pulse is applied to the discharge probe, and the output signals are detected by the capacitive sensor and displayed on an oscilloscope. The differential voltage between upper and lower electrode signals on the capacitive sensor is then measured.

The smaller the voltage differential is, the more effective the static-shielding characteristics of the test sample are. With the application of a 1,000-volt discharge pulse, nonshielded materials typically exhibit voltage differentials in excess of 300 volts. By contrast, materials with excellent static-shielding properties will exhibit voltage differentials of less than 30 volts.

Ground Strap Testers

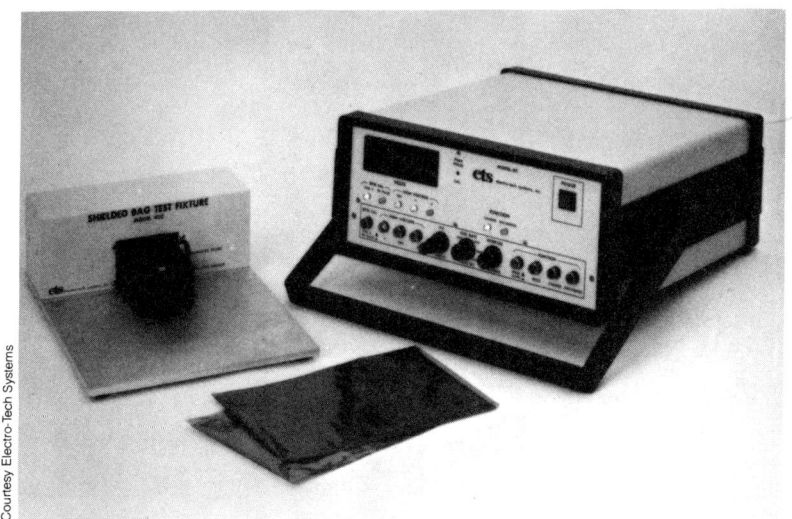

Fig. 8-13. Shielded-bag test system.

GROUND STRAP TESTERS

The wrist strap and, to a lesser degree, the heel strap are commonly used protective devices for grounding personnel during the handling and assembly of ESD-sensitive components and circuit boards. Random failure of the wrist strap can cause unsuspected damage to devices or circuit boards being handled, and create a shock hazard to the wearer.

Ground strap testers are used to check the specified resistance limits of all types of wrist straps and similar personnel grounding devices. Some of these instruments might be battery powered, they can be placed on the workbench or be carried by supervisors to rapidly check the integrity of the ground system or individual wrist straps. Some units are ac-powered wall- or pedestal-mounted.

The 250-series ground strap testers from Electro-Tech Systems (FIG. 8-14) measure the resistance of ground straps and compare the measurement to preset upper and lower resistance limits. The Model 250 wrist strap tester (center) is a portable, battery-powered unit for ESD applications in which 1-megohm wrist straps are specified. The limits are 750 kilohms and 1.25 or 10 megohms. A two-position switch on the front panel permits the selection of either a 10-megohm upper limit for checking the condition of the ground strap while attached to the wearer, or 1.25 megohms for checking the wrist strap alone. LED lamps and an audible signal indicate either a pass or a fail signal at each setting.

Electrostatic Test Equipment

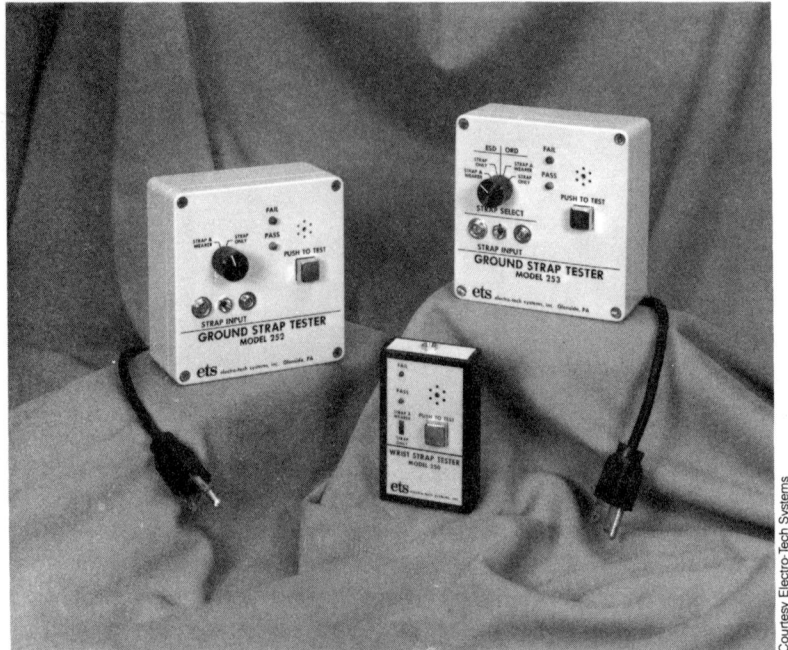

Fig. 8-14. Ground strap testers.

The Model 252 (left) is a wall- or pedestal-mounted ac-powered unit with the same features as the Model 250. The Model 253 (right) is a military version of the Model 252 that has additional settings for ground straps used when handling munitions.

9

Electrostatic Discharge Simulation Equipment

Electronics products are becoming more vulnerable to the damaging and destructive effects of ESD because the newer ICs have higher gate densities than the earlier generations of devices. The latest high-density devices are in demand because they are smaller, faster, and consume less power than the earlier components. These qualities permit end product manufacturers to provide more functions and value for the dollar, in a highly competitive industry.

Unfortunately, the narrower line widths, smaller geometries, and thinner dielectrics of these ICs make them even more vulnerable to the destructive effects of ESD than the devices in the previous generation. Protective circuits that worked in the earlier generation might not work in advanced products; this situation requires continuous research and development. Consequently, device designers are always playing catch up and they have not been able to conquer the ESD problem thus far.

Considerable progress has been made with raising the vulnerability of ICs to ESD since its destructive effects were first discovered. Through the introduction of on-chip protective circuits, the threshold of destruction of some devices has been raised from as low as 50 volts to 3,000 volts or more. Some manufacturers now claim destruction thresholds as high as 8,000 volts for their ICs.

Clamping circuits are also being added to circuit boards for additional protection or in cases in which on-chip protection has been limited. However, the devices must be handled carefully until they are soldered on the board in order for this protection to be effective. Some devices remain vulnerable, even on boards with protective circuits, because of the *antenna effect* of conductive pathways on the circuit board.

Electrostatic Discharge Simulation Equipment

The introduction of high-density silicon-gate CMOS and GaAs ICs has further complicated the protection issue which is now of major concern to the manufacturers of all semiconductor devices, products, and systems.

Both semiconductor and host equipment manufacturers must be able to measure their product's ESD sensitivity so that users will be aware of the product's degree of vulnerability and take appropriate protection measures. Determining and and reducing a product's ESD susceptibility calls for an understanding of the nature and behavior of ESD. These steps also make reliable simulation equipment essential. The simulators must be capable of reproducing ESD and its effects accurately and uniformly at various voltage and current levels.

As stated throughout this book, people are the prime source of ESD that damages electronics. Consequently, an effective simulator must be capable of simulating the charge storage and discharge characteristics of the human body. Unfortunately, there is no universal agreement on the capacitive and resistive values that best represent the human body over the wide range of conditions electronics products will encounter. Modifications to existing circuits are suggested periodically as a result of ESD research.

Specialists in the field now believe that no one human body simulation circuit will suffice for all test conditions. There are now simulator circuits that determine destructive thresholds, as well as those that determine latent damage thresholds. ESD can also be caused by static discharge from inanimate objects, and even machines that handle and package sensitive devices. Therefore, in any thorough test program, these mechanical conditions must be simulated as well.

Test methods have been developed to determine if a device, circuit board, or EUT is adequately protected. If it is not, some simulators are designed to provide diagnostic information to assist the design engineer in decreasing the product's ESD susceptibility.

Different standards are now used to test a wide variety of products—from semiconductor devices to computers and process control equipment. Currently, none of the standard ESD methods described here is acceptable for ESD testing of all products in all countries. Moreover, it is unlikely that any one standard that presently exists will be accepted for all types of testing. Biases exist for and against certain ESD test standards because of perceptions about the sponsoring agency or the test's national origins.

The U.S. Department of Defense requires that manufacturers of military specification semiconductor devices for use in U.S. military equipment test their devices with the simulation circuit specified in MILSTD-883C, Method 3015.6. This test method is discussed in chapter 2. Although the method has been updated, many experts on ESD testing

Electrostatic Discharge Simulation Equipment

are still critical of the test circuit (see FIG. 2-21). The experts believe that discharges from this circuit only approximate true static discharges, and that the circuit is not sophisticated enough to reproduce the subtleties of natural ESD.

Most ESD experts do not object to using this circuit to determine destruction thresholds of devices. However, they say the waveforms generated by this circuit did not have enough purity, and their rise times are too slow to obtain meaningful, reproducible results for testing all products. As a result, OEMs not required to use the DOD test method have elected to use other standards that they believe are more realistic. Many companies are concerned more with analyzing the causes of latent damage from ESD than determining their devices' thresholds for destruction or qualifying them for immunity.

The test circuits called out in the alternate standards are also simple RC circuits with value modifications for the human-body discharge resistance and capacitance used in the Method 3015.6 block diagram. The values selected are believed to be more realistic representations for the intended testing.

Experimental evidence has shown that discharges from hand holding a metal object such as a ring, key, or tool—known as *hand/metal* ESDs— typically have peak current amplitudes about five to seven times greater than the discharge from the hand or fingers. Investigators have concluded that some of the alternative standards are better than others because they can reproduce waves with steeper fronts and better reproduce these more realistic hand/metal discharges.

Tests conducted at various laboratories have shown that human-body ESD can have rise times of less than 200 picoseconds. Alternate circuits to the one specified in Method 3015.6 usually produce waveforms with faster rise times or steeper wavefronts. These simulator circuits are used to determine the threshold levels of certain kinds of latent ESD that cause damage.

In addition, some evidence suggests that picosecond and subnanosecond rise times on ESD spikes cause more electronic equipment disruption than nanosecond pulses. As a result, there is a trend toward the use of standards that more closely simulate these conditions when testing packaged products and systems.

According to KeyTek Instrument Corporation, real-world ESD air discharges are exceedingly complex and all aspects are not fully understood. However, they have identified five basic components or elements of these discharges that they believe disrupt equipment:

- Predischarge, corona-generated, radio frequency (rf) radiation
- Predischarge electric field (E-field)
- Discharge E-field collapse

151

Electrostatic Discharge Simulation Equipment

- Discharge magnetic field (H-field)
- Discharge current injection

The company has developed equipment capable of simulating each of these elements separately to provide what it says are more realistic and comprehensive testing and qualification procedures primarily for equipment.

ESD STANDARDS

The Internal Electrotechnical Commission (IEC) has revised publication 801-2, a standard specifically aimed at electromagnetic compatibility for industrial process measurement and control equipment. The new standard calls for an air-discharge simulator with a single human-body RC circuit using 150-picofarad and 330-ohm values rather than with the 100 picofarads and 1.5 kilohms of Method 3015.6. The IEC standard requires a maximum high-voltage power supply of 15 kilovolts for both the positive and negative outputs.

The typical waveform of the output current of the IEC ESD generator (FIG. 9-1) differs significantly from the military standard waveform (FIG. 2-

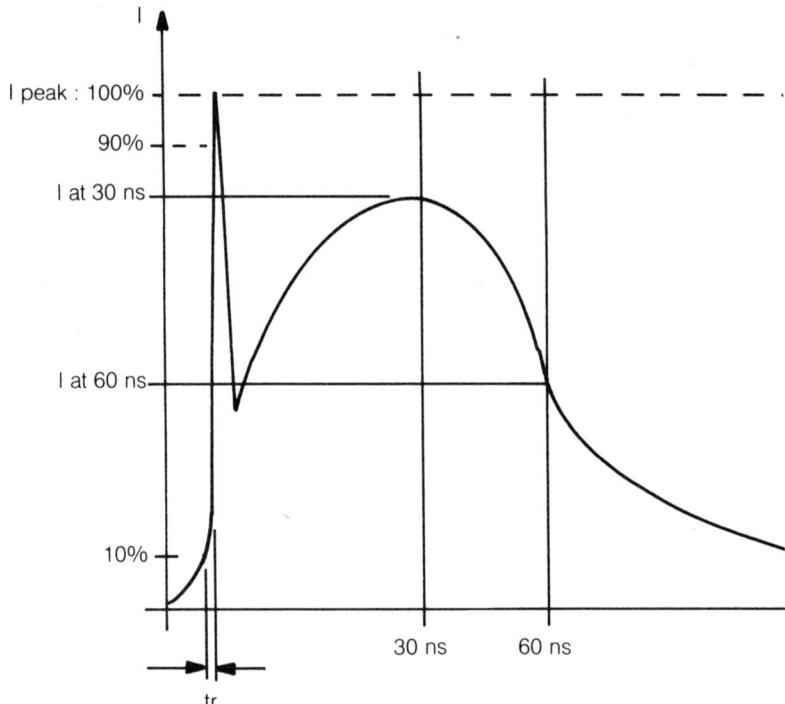

Fig. 9-1. *Typical waveform of the output current of the ESD generator per IEC 801-2.*

Commercial ESD Simulators

Table 9-1. Standards for Human Body ESD Simulation.

Agency	Standard Number	Application	Capacitance (pF)	Resistance (ohms)
DOD	MIL-STD-883C	Semiconductor devices	100	1,500
IEC	801-2	Process control	150	330
EIA	PN 1361	Voice telephone terminals	60 100	10,000 500
UL	991	Solid state controls	100	1,500
NEMA	Part DC33	Household electronic controls	100	500
SAE	J1113	Vehicle electronics	300	2,000
ECMA	TR/40	Computers	150	330

22). Many persons in the industry believe the IEC ESD generator is more realistic; its steeper leading edge spike should be noted. Rise times with the discharge switch are specified as 0.7 to 1 nanosecond compared to the less than 10 nanoseconds for Method 3015.6.

The Electronic Industries Association (EIA) is reviewing an air-discharge standard based on both 100 picofarads and 500 ohms to 10,000 volts, and 60 picofarads and 10 kilohms to 20,000 volts for voice telephone terminals. The National Electrical Manufacturers Association (NEMA) is also reviewing a standard that uses a single RC human-body circuit. Basically the same as the MIL-STD-883C circuit, it is intended for testing residential electronic controls. The NEMA standard is the first to specify separate electric and magnetic field tests.

Standards have also been written by the European Computer Manufacturer's Association (ECMA) for testing computers. The Society of Automotive Engineers (SAE) has a standard for testing vehicular electronics, and Underwriter's Laboratories (UL) has one for testing solid-state controls. The values and objectives of these standards are summarized in TABLE 9-1.

COMMERCIAL ESD SIMULATORS

Commercial ESD simulators usually include a power supply, a test circuit with replaceable or variable resistors and capacitors, and a meter for reading voltage. Most simulators are designed to reproduce the MIL-STD-883C, Method 3015.6 circuit because of its wide use in qualifying semiconductor devices. These simulators might also include provisions

Electrostatic Discharge Simulation Equipment

for altering the resistance and capacitance values to reproduce the circuit in at least one other standard.

Some commercial ESD simulators are capable of providing discharge pulses up to 30 kilovolts. These simulators are more likely to be used when testing equipment than when testing individual components. Some high-voltage simulators can generate both current-injection and spark-discharge ESD pulses. They might also include replaceable tips or probes for simulating E- and H-fields.

Bench-type MIL-STD-883 ESD Simulator

The Model 811/902 electrostatic discharge test system from Electro-Tech Systems (FIG. 9-2) simulates ESD pulses from the human body, machines, and charged device models. The system provides the discharge waveforms that meet the Class 1 ESD failure threshold—0 to 1,999 volts—specified in MIL-STD-883C, Method 3015.6

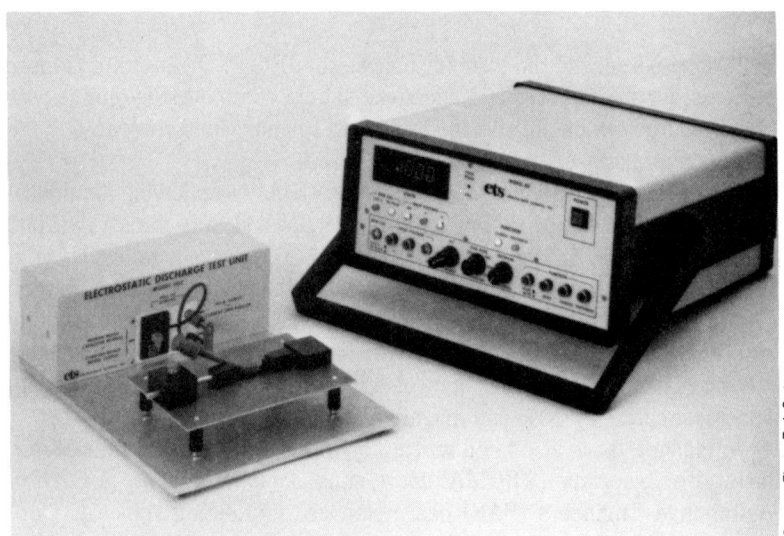

Fig. 9-2. Benchtop electrostatic discharge test system.

The Model 811 high-voltage controller, a variable high-voltage power supply and control unit, provides the necessary voltages and control signals to operate the Model 902 ESD test unit. The controller generates 0 to ±2,000 test voltages, which are displayed on an LED readout. The charging voltage rate can be selected from 50 to 2,000 volts/second in six steps, and the test system can be operated either manually or automatically.

Commercial ESD Simulators

The Model 902 test unit contains a relay that produces a waveform having a rise time that meets the requirements specified in Method 3015.6 (see FIG. 2-22). The output panel contains plug-in capacitor and resistor modules that permit selection of test configurations. The test pin of the DUT can be connected to a curve tracer or leakage tester during the charging cycle and disconnected during the discharge cycle. A series of standard and optional plug-in adaptor modules provides the necessary test system-to-DUT interface.

Hand-held and Portable ESD Simulators

Two versions of the hand-held MiniZap ESD simulator from KeyTek Instrument Corp. are available for testing computers and semiconductor-based equipment. The MZ-10 (FIG. 9-3) provides air discharge and bipolar operation up to 10 kilovolts. The MZ-15 provides current inspection in compliance with the revision to IEC standard 801.2 and is capable of a 15-kilovolt output.

Fig. 9-3. Hand-held ESD simulator is capable of a 15 kV output.

Both simulators provide air-discharge ESD simulation pulses with less than 0.3-nanosecond nominal rise times up to 5 kilovolts, independent of charge voltage. The RC networks include 150-picofarad capacitors and 330-ohm resistors, in accordance with the IEC standard. Both simulators also offer single-shot, as well as 1- and 20-per-second repetitive operations. An LED bar display measures actual high voltage at the tip,

Electrostatic Discharge Simulation Equipment

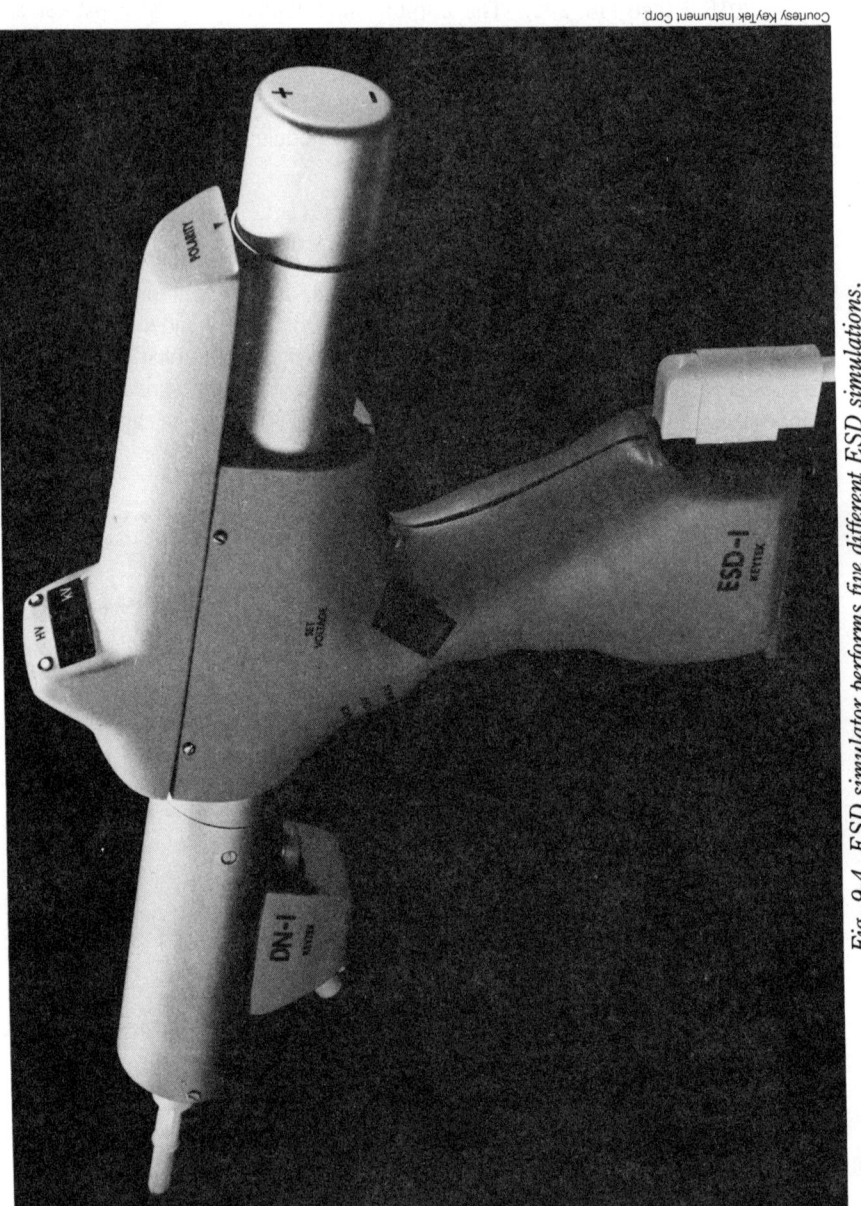

Fig. 9-4. ESD simulator performs five different ESD simulations.

Commercial ESD Simulators

and a fast-response peak indicator gives a repetition rate display. The simulators can operate with either standard line current or with 4 NiCad batteries.

The KeyTek Series 2000 is an expandable ESD simulation system intended for laboratory and production testing of computers and other electronic systems. The base unit is the ESD-1, the gun-shaped air-discharge simulator shown in FIG. 9-4. It can be hand held or mounted on a tripod. Capable of output voltages of ± 1 to ± 25 kilovolts, the ESD-1 allows the operator to select the output polarity and is capable of single-shot and repetitive operation. The ESD-1 has a built-in digital voltmeter and controls.

Accessories for the Series 2000 include a plug-in IEC human-body discharge network, plug-in discharge tips, field tips, and current-wave injection adaptors. Discharge tips, in various shapes, permit the simulation of discharges from hand-held tools and the human body. Plug-in, field-generating tips permit the simulation of four of the five basic elements of air discharge: predischarge, corona-generated rf, pre-discharge E-field, discharge E-field collapse, and discharge H-field. Adaptors permit direct current injection, the fifth element of air discharge.

Fig. 9-5. Computer-based ESD simulator qualifies up to eight ICs simultaneously.

Electrostatic Discharge Simulation Equipment

Multiple-device Tester

KeyTek has also introduced a computer-based test system for the simultaneous qualification of up to eight ICs for ESD immunity. The ZapMaster (FIG. 9-5) automatically selects and applies test impulses to ICs in selectable 50-volt steps to a maximum of 8 kilovolts. The ZapMaster also makes curve-trace measurements in 10-nanoampere increments.

The ZapMaster can test eight ICs without exceeding the industry standard maximum of one zap per second per device. The system includes a powerful personal computer to manage system functions and select test waveforms. The tester provides impulses with rise times from less than 1 to more than 10 nanoseconds. In addition to other ESD tests, it can reproduce the basic human-body model in compliance with Method 3015.6 of MIL-STD-883C with a rise time of 5 nanoseconds.

10
The Complete ESD Control Program

The previous chapters have covered the causes of ESD and its effects on sensitive electronics. They also have covered ESD-protective packaging, materials, grounding straps, clothes, equipment of an ESD-controlled workstation, ionizers, and test instruments. This chapter outlines a general program that can be adapted to all facilities where ESD-sensitive products are handled. A complete ESD control program calls for five major steps:

- Preparation of ESD-control program guides
- Design and construction of ESD-protected areas
- Preparation of ESD-control operating, handling, packaging, and marking procedures
- Development of ESD-protection training programs
- Certification of ESD-protected areas and grounded workstations

The information presented here combines many specifications and recommendations for controlling ESD and handling ESD-sensitive products. When using this information, the ESD program planner should make sure that no conflicts exist between provisions for ESD-control in contracts and the procedures stated here. Cost-effectiveness criteria can be applied to ESD control; a proper balance can be achieved between the cost of protection and the savings from reduced losses.

The Complete ESD Control Program

PREPARATION AND MONITORING OF AN ESD CONTROL PROGRAM

A successful ESD control program requires support from the top management of the company requiring this protection. If management is convinced that the company needs ESD protection, it will purchase plant improvements, including humidity controls, suitable floor surfaces, ESD-controlled workstations with personnel ground straps, and ESD-protective containers for work in process and shipment. Management would also authorize the necessary changes in operating rules to ensure the safe handling of all ESD-sensitive products and would enforce those rules.

Which level of management must decide whether to implement an ESD control program—top or intermediate—will be influenced by the cost of the program and by the number of departments participating. Quality assurance and engineering certainly will be affected, but so will manufacturing, field engineering, and other departments. Leaving any pertinent department out of the planning and implementation stages could create dissension and cause the program to fail.

Once management approves the concept of ESD control, an ESD task force or committee should be established to carry out the program. The committee should include representatives from each department concerned with the issue. They should make sure that the company's products are designed, produced, handled, and maintained in a manner that achieves the highest ESD immunity for the cost. Typical objectives for this committee are:

- Approve a design specification that states the minimum acceptable levels of product susceptibility to ESD
- Raise the employees' level of awareness concerning the impact of ESD on the company's business
- Implement manufacturing processes that control and minimize the effects of ESD

After the program is set up, it must be monitored to ensure that all personnel continue to follow the proper ESD control procedures.

The cost of constructing a completely ESD-protected room should be estimated. These estimates should be compared to the cost and administrative procedures required to establish an ESD-protected work area on an open production floor, where strict enforcement of ESD-sensitive product-handling procedures would attempt to compensate for the uncontrolled environment and personnel traffic. Management should evaluate ways to ensure that trained employees remain motivated and willing to accept responsibility for following meticulous product-handling procedures, and are consistent in their use of ESD-protective accessories and equipment.

Preparation for Work at an ESD-Controlled Workstation

Two approaches are common in the design of ESD-controlled areas: 1) consider all ESD-sensitive products highly susceptible, then standardize controls throughout the facility to meet those requirements; 2) only implement those controls needed for the category of product sensitivity within each specific protected area. These controls could be based on device sensitivity in accordance with circuit test Method 3015.6 MIL-STD-883C.

PREPARATION FOR WORK AT AN ESD-CONTROLLED WORKSTATION

The cornerstone of a successful ESD control program is the protected workstation connected to ground through a 1-megohm ± 10 percent resistor, a grounding wrist strap with the ground wire connected to the static-dissipative work surface, an ionizer, and humidity controls. Also recommended are a wrist-strap continuity and resistance checker, grounded tools and test equipment, and ESD-dissipative containers for work in progress. TABLE 10-1 lists the specifications for an ideal ESD protective area and FIG. 10-1 shows a typical ESD-controlled workstation. The following procedures and precautions are recommended:

- The wrist strap must be in snug contact with bare skin, usually on the left wrist of a right-handed person. The wrist strap must be worn the entire time an operator is at an ESD-controlled station.

- Persons should first touch the grounded benchtop before handling ESD-sensitive items. This precaution should be observed, in addition to use of a grounding wrist strap. If possible, operators should avoid touching leads or contacts, even though they are grounded.

- An operator's clothing should never make contact with or be close to ESD-sensitive items. Operators must be especially careful to prevent static-sensitive items from touching clothing. If static-dissipative smocks are not used, long sleeves must be rolled up or covered with antistatic sleeve protectors that "cage" the sleeve at least up to the elbow. If static-dissipative smocks are used, they must completely cover long sleeves. The cleaning instructions for the smock must be followed.

- Only cotton gloves or antistatic finger cots, free of reactive elements such as chlorine or phosphorous, must be used when handling ESD-sensitive items.

- Any person not properly prepared, as just outlined while at or near the workstation must not touch or come close to ESD-sensitive products.

- It is the responsibility of the operator and the supervisor to ensure that the static-free work area contains no static hazards, including such personal items as plastic-coated cups or wrappers, plastic cosmetic bottles

The Complete ESD Control Program

Table 10-1. Model Specification for ESD Control.

Item		Remarks
1. Work Area		
a) Size: 2,000 – 3,000 sq. ft.	i	Accommodate 30 – 50 people.
	ii	Limit the number of exits and entrances, hence environmental conditions are controllable.
b) Floors: Conductive tiles	i	Easier to maintain and clean.
c) Ceiling	i	Necessary for stability of environmental control.
d) Lighting level	i	300 ft. candles.
e) Ventilation	i	Provides for air circulation
2. Environmental Controls		
a) Humidity: 45% ± 5%	i	Comfortable and pleasant.
b) Temperature: 68°F ± 5°F	i	Heat generating equipment will raise the temperature to 76°F (24.4°C).
	ii	Limits rusting and corrosion. Reduces ESD arc generation.
c) Smoking	i	Smoking should not be allowed.
	ii	Smoke particles can form discharge paths or deposit charges when they settle on assemblies in work areas.
	iii	Nicotine and other contaminants create part solderability and conformal coating problems
d) Food and beverages	i	No eating or drinking is to be allowed.
	ii	Will encourage bacterial or fungal growth, contaminate PC boards and assemblies.

Preparation for Work at an ESD-Controlled Workstation

3. Benchtop Materials

   a) Surface resistivity range $10^6 - 10^9$ ohms per square
      i   Material manufacturers should supply data sheets that specify resistivities at RH levels of 30% – 60% and temperatures of 60°F – 80°F.
      ii Very fast decay times generate secondary RF fields that can destroy sensitive devices via discharge to ground.

   b) Static decay times 0.1 – 0.2 seconds from 5000 volts.

   c) Chemical properties
      iii Must not support bacterial or fungal growth.
      iv Must not form ionic layers with PC boards to release particles and other contaminants from solderable surface.
      v   Shall be nonflammable and nontoxic by oral injection, inhalation or dermal application.
      vi Should provide some cushioning and must not depend on humidity to function.

4. Protective Clothing

   a) Coveralls, labcoats and smocks
      i   Cotton labcoats are comfortable.
      ii Maintenance of low residual ESD voltage on synthetic labcoats is almost impossible.
      iii Labcoats must not be worn outside static free work areas (i.e., washrooms, cafeterias, and outdoors, etc.).

   b) Wrist straps
      i   Mandatory

   c) Conductive shoes
      i   Necessary in microcircuits labs or ESD-protected areas with conductive floor.

163

The Complete ESD Control Program

Table 10-1 cont.

Item	Remarks
5. Protective Material	
a) Tote boxes, containers	i Should have the same properties as per paragraph 3 of Model Specification.
b) Transport carriers	
c) Document holders	
6. Ionizers	i Use with caution, especially in conjunction with chemicals such as conformal coating areas.
7. Cleaning agents	
Antistatic sprays and solutions	i Must have same specifications as per paragraph 3c.
	ii Must not be corrosive.
	iii For microcircuits labs and fabrication areas, only solutions free of acid radicals, sodium, phosphates, chlorides and sulphates-free solutions must be used.

Preparation for Work at an ESD-Controlled Workstation

or boxes, combs, tissue boxes, cigarette packages, and vinyl or plastic purses. All work-related items including instruction sheets, fluid containers, tools, and parts containers must be approved for use in the ESD-controlled area (see TABLE 2-2). In addition, procedures must be established to ensure that these objects are not brought into the room by personnel not trained to work in the controlled area.

Fig. 10-1. ESD-controlled workstation.

The Complete ESD Control Program

- A minimum level of protection can be established in the field for basic maintenance by a person with a knowledge of the principles of ESD protection. It could consist of an area kept free of static generators and equipped with a personnel wrist ground strap, a portable protective work mat, ESD-protective containers for repair parts and spares, and ESD-protected hand tools. A typical portable field service kit is shown in FIG. 7-7.
- Protected work areas, workbenches, and containers should be identified with ESD caution signs.
- Access to ESD-protected areas should be restricted to persons who have been properly trained in ESD control procedures, have official business in the area, and are dressed in clothing suitable for the protection level in force. Visitors not meeting all of these qualifications should be escorted by a responsible member of management who will advise them on protective procedures and warn them not to touch any ESD-sensitive devices or products.
- Persons handling ESD-sensitive components should avoid physical activities that produce static in the vicinity of sensitive products. These activities include wiping feet, removing or putting on smocks, or moving furniture and other heavy objects.

GENERAL HANDLING PROCEDURES AND REQUIREMENTS

The following general procedures and precautions are recommended throughout the ESD-controlled area:

- All ESD-sensitive items must be received in a closed antistatic/conductive container and must not be removed from the container, except at a static-free workstation. All protective folders or envelopes containing documentation (lot travelers, etc.) must be made of a material that does not accumulate static.
- Each packing (outermost) container and package (internal or intermediate) must have a brightly colored warning label attached, stating the following information or equivalent:

CAUTION!
ELECTROSTATIC
SENSITIVE
DEVICES
Do Not Open or Handle
Except at a
Static-Free Workstation

General Handling Procedures and Requirements

The warning label must be legible with normal vision at a distance of 3 feet.

- ESD-sensitive items should remain in their protective containers, except when they are being worked on in an ESD-controlled area.
- Before removing the items from their protective container, the operator should:
    ○ Place the container on a static-dissipative surface.
    ○ Make sure the wrist strap fits snugly and is electrically connected to the ground receptacle.
    ○ Touch the static-dissipative surface.
- All work should be performed with the items in contact with the static-dissipative work surface, when possible. Conductive DIP tubes or magazines should not touch hard-grounded test equipment on the bench.
- When it is impossible or impractical to ground the operator with a wrist strap, a conductive shoe strap can be used with conductive flooring or floor mats.
- When the operator moves from any other place to the ESD-controlled workstation, the start-up procedure must be the same as in the previous section.
- If an ionizer is used, it must be in operation before ESD-sensitive items are removed from their packaging, as well as during the entire period the items are at the station.
- ESD-sensitive items must never be allowed to be sent in polystyrene foam *shells* or *peanuts*, or other high-dielectric materials, unless they have been treated with an antistat. This treatment should be verified by noting a change in color and measurement of the material's inability to generate more than ± 100 volts. If the treated materials are broken, they must be discarded.
- ESD-sensitive items must not be transported or stored in trays, tote boxes, vials, or similar containers made of an untreated plastic material, unless the items are protectively packaged in conductive material.

GENERAL GUIDELINES FOR HANDLING ESD-SENSITIVE DEVICES

The following general guidelines apply to the handling of ESD-sensitive devices, particularly MOS and GaAs ICs. All recommendations from the previous sections apply.

- The maximum ratings specified by the data sheet must not be exceeded.

The Complete ESD Control Program

- All unused device inputs must be connected to power supply or ground leads, whichever is appropriate to the specific device.
- All low-impedance test equipment (pulse generators, etc.) must be connected to ESD-sensitive device inputs only after the device is powered up. Similarly, this equipment must be disconnected before power is turned off.
- All ESD-sensitive devices must be stored or transported in containers appropriate to their susceptibility to ESD damage. Category A devices, per MIL-STD-883, with an ESD sensitivity of 20 to 2,000 volts, must be stored in conductive or antistatic containers within an electrostatic field shielding barrier. Category B devices with an ESD sensitivity greater than 2,000 volts must be stored in antistatic containers. These devices must not be inserted into untreated polystyrene foam or plastic trays, but must be left in their original containers until ready for use.
- All ESD-sensitive devices must be placed on a grounded bench or table surface, and operators must ground themselves before handling such devices.
- If automatic handling machines are used, high levels of static electricity can be generated by the movement of the device, the belts, or other moving parts. The static buildup should be reduced by using ionized air blowers or room humidifiers. All parts of the machine that come in contact with the top, bottom, or sides of IC packages must be grounded to metal or other conductive materials.
- Cold chambers, using carbon dioxide for cooling, should be equipped with baffles, and the ESD-sensitive devices must be contained in or placed on conductive materials.
- When straightening device leads or hand-soldering is necessary, ground straps must be provided for the tools used and the soldering ties should be grounded.
- The following procedures must be observed during wave soldering:
  - The solder pot and conductive conveyer system of the wave soldering machine must be hard-grounded.
  - The work surfaces for loading and unloading must have conductive tops and be hard-grounded.
  - Completed circuit boards and subassemblies must be placed in antistatic containers before being moved to the next stations.
- The use of static-detection meters for surveillance of the protected area is highly recommended.
- ESD-sensitive devices must not be inserted or removed from test fixtures with the power applied. All power supplies used for testing devices must be checked to make certain that no voltage transients are present.

General Guidelines for Handling ESD-Sensitive Circuit Boards

- The test equipment set-up must be rechecked for proper polarity of power and ground terminals before parametric or functional testing is conducted.
- DIP magazines must not be recycled; repeated use causes their antistatic coating to deteriorated.

GENERAL GUIDELINES FOR HANDLING ESD-SENSITIVE CIRCUIT BOARDS

The following general guidelines apply to the handling of loaded circuit boards. All recommendations from the previous sections apply.

- Circuit boards containing ESD-sensitive devices are extensions of those components; the same handling procedures apply for both. Touching edge connectors wired directly to device inputs can cause damage. Plastic wrappings must be avoided. When external connections to a circuit board are connected to an input of an ESD-sensitive device, a resistor should be used in series with the input. This resistor helps limit accidental damage if the circuit board is removed and brought in contact with static-generating materials. The limiting factor for the series resistor is the added time delay.
- Rigid conductive shunts, made as board edge clips, must be used to short the terminals of ESD-sensitive circuit boards that are not being tested or inspected. The circuit boards should be placed in ESD-protective bags or boxes.
- Shorting bars, clips, or noncorrosive conductive foam must be used to cover connector terminals of circuit boards and higher-level assemblies. Conductive outer wrappings should be used when an ESD-sensitive product is being moved from ESD-protected areas.
- The following steps must be observed during board-cleaning operations:
  - Vapor degreasers and baskets must be hard-grounded.
  - Brush or spray cleaners must not be used.
  - Assemblies must be placed into the vapor degreaser immediately upon removal from the conductive or antistatic container.
  - Cleaned assemblies must be placed in antistatic containers immediately after removal from the cleaning basket.
  - The application of high-velocity air should occur only when assembled circuit boards are grounded and the output of an air ionizer is directed at the board.
- Tools and test equipment used in ESD-controlled areas must be grounded properly. Hand tools should not have insulated handles; however, if they do, the insulated handles must be treated with an antistat.

The Complete ESD Control Program

- Circuit boards and subassemblies must not be energized while ESD-sensitive components, especially those containing MOS or GaAs transistors or ICs, are being removed or inserted.
- All power supplies used for testing ESD-sensitive circuit boards must be checked to make sure that no voltage transients are present.
- The test equipment setup must be checked for proper voltage polarity before parametric or functional testing is conducted.
- Test specifications and diagrams for ESD-sensitive circuit boards must be marked with the ESD-sensitive electronic symbol, warnings, precautions, and handling procedures, as applicable.
- Manufacture, process, assembly, and inspection work instructions must identify ESD-sensitive circuit bords for ESD control. These instructions should require that these items be handled outside of ESD-protective containers only by trained persons within ESD-controlled areas.

RECEIVING INSPECTION PROCEDURES FOR ESD-SENSITIVE ITEMS

The preceding guidelines are valid for all areas in a plant where ESD-sensitive products are handled. The following procedures are needed for receiving inspection and stockroom operations.

- Containers of ESD-sensitive items must not be accepted into the stock unless they are identified as containing ESD-sensitive items.
- Items must be removed from the protective container for order subdividing by a properly trained and grounded person at an approved ESD-controlled workstation.
- All subdivided lots must be carefully repackaged in ESD-protective containers before they are taken from the ESD-controlled workstation. The containers must be labeled to indicate that the packages contain ESD-sensitive items. If it is suspected that the ESD-sensitive item is not adequately protected, it should not be transferred to another container. The originator should be contacted and an appropriate disposition should be negotiated.
- It is the responsibility of the supervisor to make sure that all personnel assigned to this operation are familiar with the handling procedures (outlined in the previous sections). Everyone who moves stock should be instructed to avoid direct contact with unprotected ESD-sensitive items.
- ESD-sensitive items must not be accepted into the packing area unless they are in an ESD-protected bag or conductive container.

Field Testing and Inspection

- An ESD-sensitive item that is delivered within an approved container and is found to be properly identified must be repacked in ESD-protective packaging, in a standard shipping carton or other regular packaging material. Containers must be labeled in accordance with the general handling procedures.
- Any void filler must be made of antistatic materials.

FIELD TESTING AND INSPECTION

The following procedures apply to testing in the field. All of the recommendations from the previous sections apply.

- ESD-sensitive products should be tested in ESD-protected areas at grounded workstations to the extent practical. If not, grounding mats and personal ground straps approved for field work must be used.
- Diagnostics must be performed to isolate the faulty assembly. Do not use pressurized aerosol coolant in fault isolation.
- Equipment power must be shut off.
- Prior to touching any ESD-sensitive products, personnel must put on personal wrist strap and connect the other end of the ground cable to the equipment cabinet or chassis ground. Where ground straps cannot be used, personnel should touch the grounded equipment chassis with their hands prior to removing or inserting a sensitive item.
- After the failed ESD-sensitive component or circuit board is removed, it must be placed in an ESD-protective bag or other appropriate container pending further testing. Perhaps a complete SEM analysis to determine the precise mechanism for damage or destruction will be necessary.
- ESD-sensitive products should not be tested with test equipment leads unless necessary. Ideally, all test procedures should be done only under ESD-controlled conditions. When probing is necessary, the meter and probes or test leads must be grounded prior to touching the terminals of the sensitive items.
- The package should be opened at the connector end if possible. Then the ESD-sensitive product should be removed from the ESD-protective packaging and installed in the equipment. Touching of adjacent parts, electrical terminals, and circuitry must be avoided.
- All other required final maintenance adjustments, such as tightening of fasteners and replacing covers, should be performed before the personnel ground strap is removed. If a ground strap is not or cannot be worn, the grounded equipment chassis should be touched prior to each action.
- Equipment can then be energized and verified for fault correction.

The Complete ESD Control Program

AUDIT PROVISIONS FOR ESD CONTROL

An ESD-controlled area will remain effective only by continuous compliance with the specification in force for that area. Regular testing procedures and deficiency correction is a requirement. The following recommendations are made for an ESD-controlled facility:

- Each plant location in which ESD-sensitive products are handled must be audited at least once each quarter for compliance with all terms of the applicable internal or contract specifications. Ground continuity and the presence of any uncontrolled static voltages in excess of ± 100 volts are considered critical. If continuity is broken and higher voltages are detected, checking must be done more frequently.
- Ground continuity should be checked at least once per week on wrist straps and ground wires. The presence of 1-megohm ± 10 percent resistors in the ground connection, between the work surface and both personnel wrist straps and ground must be verified.
- Grounded conditions must be checked at least once per week. A visual inspection must be made to determine that the equipment fully complies with specifications at ESD-controlled workstations during the handling of ESD-sensitive components. These specifications cover proper operator grounding, handling of ESD-sensitive items in ESD-controlled areas only, and the absence of all static-generating materials at ESD-controlled workstations.
- Sleeve protectors must be checked at least once per week. A visual check must be made to verify that each operator wearing loose-fitting or long-sleeved clothing either has his or her sleeves properly rolled up or is covered with sleeve protectors properly grounded to the bare skin at the wrist. Sleeve protectors are not required if static-dissipative smocks or shopcoats are worn.
- Static voltage levels must be checked at least once per week. In addition to visual inspections, electrostatic voltmeters must be used to check for uncontrolled electrostatic voltages exceeding ± 100 volts at or near ESD-controlled workstations. If static-dissipative smocks are worn, they must also be checked to the same level.
- Conductive floor tiles must be checked at least once per month. Conductive flooring must have a resistance of at least 25 kilohms from any point on the tile to hard ground. Also, resistance between any two points on the tile flooring, 3 feet apart, must be at least 25 kilohms. The test methods used should be ASTM-F-150-72 and NFPA 56.
- Conductive floor mats must be checked at least once per week. The ground connection, through a 1-megohm ± 10 percent safety resistor,

Personnel Training and Certification

must be checked for continuity. The mat must be clean and free of tears or holes.
- If a condition is found that is not in compliance with specifications, the processing must be stopped until the condition is corrected.
- Written records must be kept of all tests performed.

PERSONNEL TRAINING AND CERTIFICATION

Responsible, motivated and properly trained personnel are absolutely necessary for an effective ESD control program to be carried out successfully. Facilities with the finest test instruments, best equipment, and complete provisions for safeguarding ESD-sensitive products cannot prevent the damage or destruction of products by untrained, careless, or indifferent supervisors and employees.

Persons assigned to work within an ESD-controlled area can be motivated to carry out all procedures and learn proper work habits if they are well informed about the causes of ESD and its consequences in damaged or destroyed products. In addition, because of the financial loss to the company, the jobs of all employees could be at stake if those losses are not controlled.

Personal responsibility begins with an awareness of the ever-present invisible, imperceptible threat of ESD and extends to the importance of wearing ground straps. It also includes a consciousness of proper clothing and the importance of keeping personal papers, plastic and paper cups, and other potential static sources out of the protected work area. Many organizations have found that this behavior is best learned with formal training programs that include lectures, textbooks, notebooks, and the keeping of written notes. These programs are reinforced by films, video tapes, and perhaps even demonstrations of actual ESD destruction of devices.

Training in static awareness and ESD-protective procedures should be given to all people who specify, procure, design, or handle ESD-sensitive products. Supervisors should be aware that certain instructions given to employees might contradict general ESD-protection guidelines, particularly in situations when the management orders increased production.

Training should also be given to all those who have access to ESD-controlled rooms, but are not regularly stationed in them—for example, those performing janitorial and plant maintenance services. These employees must be instructed not to touch or handle sensitive components, subassemblies, or other work in progress that might be exposed in the controlled area.

ESD-control training programs should be oriented to the manufacturer's facilities and to the types of ESD-preventive facilities, tools, and equipment that are effective for the type of activity being conducted there. Personnel should be trained to use the ESD-protective bags, tote boxes, and other containers, as well as how to connect and test ground straps. The employees should understand the theories behind ESD precautions carried out during the handling of ESD-sensitive products.

ESD awareness should also be a part of training courses, instruction manuals, and application notices prepared by the original equipment manufacturer for users. This information should include identification of ESD-sensitive components in the equipment, basic ESD theory, sensitive product-handling precautions, and the need for and use of various types of ESD-protective containers and packages. It should also cover safety precautions when grounding tools, persons, test instruments, and work surfaces.

Regularly scheduled training programs should be mandatory for all new personnel. The thoroughness of training should depend on the trainee's ability and need to comprehend the information provided, and the economic consequences of negligence for which that person is responsible.

Engineers who design protective circuitry obviously need more theoretical training than stockroom personnel who place ESD-sensitive items in kits. Certificates of satisfactory completion should be given to those who have attended the training course and demonstrated their comprehension of its content with verbal or written tests. Periodic refresher training of all personnel is recommended every 12 months to maintain proficiency. This is particularly important if new and more stringent procedures are introduced.

PRODUCT DESIGN

When designing ESD immunity into sensitive products, it is important that designers consider where the end product will be used and what level of training or experience the users will have. For example, continuous procedures to safeguard ESD-sensitive subassemblies and circuit boards are easier to implement in computer rooms and laboratories than on unrestricted factory floors. Therefore, equipment destined for this environment can have a lower ESD immunity than equipment used in industrial or consumer applications. The cost of including ESD protection is influenced by low selling prices and narrow profit margins.

The type of enclosure of the equipment or instrument has a direct influence on ESD protection. Metal enclosures with complete conductive envelopes at all openings extending to cables and connectors are effective in preventing faults from ESD. If metal enclosures are not used, the application of conductive coating within nonmetal enclosures has also been effective. Cables leaving the equipment should be shielded and the shield should be connected to the chassis ground.

Customer Responsibilities

The manufacture of an ESD-protected product requires close cooperation among all engineering disciplines on the design staff. In many cases, provisions for the prevention of (EMI) will also be helpful in preventing ESD faults, and where possible, the design provisions for each should be combined.

CUSTOMER RESPONSIBILITIES

Ideally, product customers should be able to ignore ESD problems, but until all danger of ESD casualties has been eliminated, customers must share the responsibility of preventing ESD damage. Customers can help prevent damage only if they are made aware of any possibilities of ESD damage in product operation and maintenance manuals, catalogs, and applications notes.

A
Standards

U.S. GOVERNMENT DOCUMENTS

NOTE: Unless otherwise stated, the latest revision in effect governs.

FED-STD-1018 Preservation, Packaging and Packaging Materials Test Procedures; Test Method 4046: Electrostatic Properties of Materials

PPP-C-1842 Cushioning Material, Plastic, Open Cell (For Packaging Applications)

DOD-HDBK-263 Electrostatic Discharge Control Handbook for Electrical and Electronic Parts, Assemblies and Equipment

DOD-STD-1686 Electrostatic, Discharge Control Program for Electrical and Electronic Parts, Assemblies and Equipment (Excluding Electrically Initiated Explosive Devices)

MIL-STD-129 Marking for Shipment and Storage

MIL-STD-883 Test Methods and Procedures for Microelectronics: Test Method 3015: Electrostatic Discharge Sensitivity

MS-90363 Box, Fiberboard, with Cushioning for Special Minimum Cube Storage and Limited Reuse Applications

MIL-B-117 Bags, Sleeves and Tubing, Interior Packing

MIL-S-19491 Semiconductor Devices, Packaging of

MIL-M-38510 Microcircuits, General Specification for

MIL-M-55565 Microcircuits, Packaging of

MIL-B-81705 Barrier Materials, Flexible, Electrostatic-Protective, Heat Sealable

MIL-P-81997 Pouches, Cushioned, Flexible, Electrostatic-Free Transparent

MIL-W-80 Transparent Antielectrostatic Acrylic Base Observation Window

NAVSEA SE 003-AA-TRN-010 Electrostatic Discharge Training Manual

Standards

INDUSTRY DOCUMENTS

AATCC Test Method 134-1975 Electrostatic Propensity of Carpets
ANSI Test Method 47 (Secretariat) **707** (Proposed) Electronic Devices Sensitive to Electrostatic Discharges
ANSI Standard 241.3-1976 Conductive Safety-Toe Footwear
ASTM Test Method D-257-78 D-C Resistance or Conductance of Insulating Materials
ASTM Test Method D-991-85 Rubber Property-Volume Resistivity of Electrically Conductive and Antistatic Products
ASTM Test Method D-2679 Electrostatic Charge
ASTM F-150
ASTM Test Method D-3509 Electrostatic Strength Due to Surface Charges
EIA Standard RS-471 Attention Symbol and Label for Electrostatic Sensitive Devices
EIA Standard 541
EIA IS-5A Packaging Material Standards for ESD Sensitive Items.
EOS/ESD Standard No. 3
EOS/ESD Standard No. 4
NAS 853 Field Force, Protection for
NFPA Standard 56A Inhalation Anesthetics; Section 46: Reduction in Electrostatic Hazard

INDUSTRY ORGANIZATIONS

EOS/ESD Association, Inc.
P.O. Box 298
Westmoreland, NY 13490

Reliability Analysis Center RADC/RAC
Griffiss Air Force Base, NY 13441

B
Directory of Suppliers

This directory lists ESD control product manufacturers and vendors in North America. The directory contains the names of both large and small suppliers. No distinction is made between those firms that manufacture their own products and those that resell or distribute products made by others. Some of the companies listed combine both of these functions. The directory is, however, limited to suppliers of products; the names of companies that only offer consultation or special services have not been included.

The ESD-protection industry is undergoing continuous change. New firms are entering and others are leaving. Keeping track of companies is complicated because of the name changes that result from acquisitions or mergers. Some name changes reflect altered business plans. Addresses are changed as companies grow out of existing quarters. These dynamics make it difficult to prepare a directory that is both comprehensive and completely accurate for more than a few months.

ESD-production product vendors range from small, privately owned firms that either manufacture specialty items or resell goods made by others to divisions of large corporations. Evidence suggests that only a few of the hundreds of market participants hold the dominant share. Perhaps as many as 25 companies have achieved national recognition status. Most of them combine manufacture with relabeling and reselling to achieve complete ESD-control product lines.

Some of the best known companies have established their reputations as manufacturers of ESD specialties such as ionizers, shielded bags, static-dissipative containers, or test instruments.

Many companies listed in this directory specialize in apparel, flooring, containers, packaging, and tools modified to meet ESD-control requirements. Because the functions of so many ESD-protection products

Directory of Suppliers

depend on specially prepared materials or chemicals, companies with established reputations in plastics and materials technology are prominent in the field.

The names of the companies in this directory are listed with codes to represent product offerings in nine different categories:

1. **Antistats** chemicals applied to materials to increase their conductivity
2. **Clothing** outer garments including smocks, shop coats, shoes, and gloves
3. **Flooring/Work Surfaces** sheet or tile flooring and sheet or mat work surfaces suitable for an ESD-controlled area
4. **Grounding Products** wrist and foot grounding straps, cables, cords, terminals, and accessories
5. **Ionizers** active and passive machines for air ionization
6. **Materials** conductive, dissipative, or antistatic materials in sheet or bulk form—for other than flooring or work surface use.
7. **Packaging** bags, DIP magazines, trays and tote boxes.
8. **Test Instruments** instruments for measuring ESD-related variables including resistance, resistivity, and static charge as well as for ESD waveform simulation.
9. **Workstation Products** furniture, tools, and accessories modified for use in ESD-controlled areas.

```
A = Antistats              C = Clothing
F = Flooring               G = Grounding Products
I = Ionizers               M = Materials
P = Packaging              T = Test instruments
W = Workstation Products
```

A & J Manufacturing Co.   P
11121 Hindry Avenue
Los Angeles, CA 90045
(213) 678-3053

Abbey Plastics Corp.   A, G, M, P, W
11 E. Brent Drive
Hudson, MA 01749
(508) 562-6971

Abigail Uniforms, Inc.   C, P
1920 University Avenue
St. Paul, MN 55104
(612) 645-5809

Appendix B

Accutech Static Co. A, C, G, P, W
P.O. Box 70016
Pasadena, CA 91117
(818) 286-7485

ACL, Inc./Staticide A, T, W
1960 E. Devon Avenue
Elk Grove Vill., IL 60007
(312) 981-9212

ADE, Inc. M, P, W
1430 E. 130th Street
Chicago, IL 60633
(312) 646-3400

Advanced D-Stat A, C, F, G, I, M, P, T, W
31067 San Clemente Avenue
Hayward, CA 94544
(415) 489-2115

Advanced Dynamics Corp. M
2000 E. Lamar Boulevard
MD600
Arlington, TX 76006
(817) 274-2666

Advanced Web Products, Inc. A, P
522 5th Avenue
New York, NY 10036
(212) 391-1620

Advance Engineering F, G, P, T, W
777 Front Street
Burbank, CA 91503
(818) 841-1519

Aesops Inc. A, C, F, G, I, M, P, T, W
636 S. Glenwood Place
Dalton, GA 30721
(800) 235-8817

Airmold/W. R. Grace & Co. P, W
P.O. Box 610
Roanoke Rapids, NC 27870
(800) 344-5716

Directory of Suppliers

Ajusto Equipment Co. W
P.O. Box 348
Bowling Green, OH 43402
(419) 823-1861

Akzo Engineering Plastics A, M
P.O. Box 3333
Evansville, IN 47732
(812) 424-3831

Alacra Systems M, P, W
Race & Ridge Streets
Ambler, PA 19002
(800) 448-BINS

All Spec Packaging A, C, F, G, I, M, P, T, W
61-10 Maurice Avenue
Maspeth, NY 11378
(516) 868-9000

Alpha Modular/Systems, Inc. P
290 W. Hamilton Avenue
Campbell, CA 95008
(408) 378-9711

Alphastat A, G, I, M, P, T, W
28 Goodhue Street
Salem, MA 01970
(508) 744-1100

ALX Technical Service Ltd. A, F, G, M, W
320 Mill Street
Richmond Hill
Ontario, Canada L4C 4B5
(416) 884-1246

American Environmental I, T, W
Systems
4699 Nautilus Court S.
Boulder, CO 80301
(303) 530-7077

American Seating Company W
901 Broadway NW
Grand Rapids, MI 49504
(616) 732-6600

Appendix B

Ametek, Inc. M
Brandywine Four Bldg.
Routes 1 and 202
Chadds Ford, PA 19317
(215) 358-1180

Amstat Industries Inc. A, F, G, I, T, W
3012 North Lake Terrace
Glenview, IL 60025-1335
(708) 998-6210

Anderson Effects, Inc. A, G, I, M, P, T, W
P.O. Box 657
Mentone, CA 92359-0657
(714) 794-3792

Antistatic Industries Inc. C, G, I, M, P, W
130 Gamewell Street
Hackensack, NJ 07602-0460
(201) 489-7654

Apache Mills, Inc. A, F, M, W
417 River Street
Calhoun, GA 30701
(800) 241-4490

Arc Associates A, C, F, G, I, M, P, T, W
11 Henry Avenue
Feasterville, PA 19047
(215) 364-1049

Associated Bag Company A, M, P
150 S. 2nd Street
Milwaukee, WI 53204
(414) 272-2380

Astec Company A, F, G, I, M, P, T, W
P.O. Box 9
Hawthorne, NJ 07507
(201) 595-7001

Atrix Inc. A, C, G, P, T, W
14301 Ewing Avenue, S
Burnsville, MN 55337
(800) 222-6154

Directory of Suppliers

Baron Industries Inc. A, M, P
4721 Ironton Street
Denver, CO 80239
(303) 375-0770

Baxter Industrial Div. G, M, P, W
27200 N. Tourney Road
Valencia, CA 91355
(805) 253-7466

Baxter Scientific Products A, C, G, I, M, P, T, W
1430 Waukegan Road
McGaw Park, IL 60085
(312) 578-4269

Bekaert Corporation C, M
1395 Marietta Parkway
Marietta, GA 30067
(404) 421-8520

Bradford Company A, M, P, W
13500 Quincy Street
Holland, MI 49422-1199
(616) 399-3000

Brookdale Plastics G, P
1599 8th Street, SE
Minneapolis, MN 55414
(612) 331-7092

Buckhorn Inc. P
55 W. Technecenter Drive
Milford, OH 45150
(800) 543-4454

Butcher Company, The A, F
120 Bartlett St.
Marlboro, MA 01752
(508) 481-5700

Bystat Inc. A, C, F, G, I, M, P, T, W
9700 Henri-Bourassa
St. Laurent, Quebec,
Canada H4S 1R5
(514) 333-8880

Appendix B

Catalina Plastics Suppliers　　A, M, P
9830 San Fernando Road
Sun Valley, CA 91352
(818) 504-2800

Central Container Corp.　　A, G, I, M, P, T, W
4041 Hiawatha Avenue
Minneapolis, MN 55406
(612) 721-6224

Chapman Corporation　　A, I, T
125 Presumpscot Street
Portland, ME 04104-6100
(207) 773-4726

Charleswater Products, Inc.　　A, F, G, I, M, P, T, W
93 Border Street
West Newton, MA 02165
(617) 964-8370

Claus Cutlery Co.　　W
223 N. Prospect Street
Fremont, OH 43420
(419) 332-7344

Clean Room Products, Inc.　　A, C, M, W
1800 Ocean Avenue
Ronkonkomo, NY 11779
(516) 588-7000

Coatings/Composites　　A, F, G, W
10105 Doty Avenue
Inglewood, CA 90303
(800) 421-5418

Cohen Co., Arnold David　　A, C, F, G, I, M, P, T, W
261 Old Billerica Road
Bedford, MA 01730
(617) 275-2646

Colvin Packaging Products　　A, G, M, P, W
1391 Hundley Street
Anaheim, CA 92806
(714) 630-3850

Directory of Suppliers

Conductive Containers, Inc. A, G, M, P, W
425 Huehl Road, Building 2
Northbrook, IL 60062
(708) 480-7887

Contact East, Inc. A, C, F, G, I, M, P, T, W
335 Willow Street S
North Andover, MA 01845
(508) 682-2000

Coroplast M, P
3025 Skyway Circle N.
Irving, TX 75038
(214) 256-2241

Correct Products, Inc. A, C, F, G, I, M, P, T, W
1701 N. Greenville Avenue
Richardson, TX 75081
(214) 690-6645

Crystal-X Corp. A, F, M, P, W
100 Pine Street
Darby, PA 19023-3198
(215) 586-3200

Cybergen Systems Corp. I
2070 Walsh Avenue
Santa Clara, CA 95050
(408) 727-6766

Dacobas, Inc. G, W
1890 N. Voyager Avenue
Simi Valley, CA 93063
(805) 526-7733

Desco Industries, Inc. F, G, I, W
761 Penarth Avenue
Walnut, CA 91789
(714) 598-2753

Diversi-Plast Products A, P
7425 Laurel Avenue
Minneapolis, MN 55426
(612) 540-9700

185

Appendix B

Dixon Paper Co.
680 Garden of the Gods Road
Colorado Springs, CO 80907
(303) 598-5820

A, C, F, G, I, M, P, T, W

Donray Company
500 SOM Center Road
Cleveland, OH 44143
(216) 449-6450

A, C, F, G, I, M, P, T, W

Dow Chemical Co.
P.O. Box 1206
Midland, MI 48674
(800) 258-2436

P

Duracote Corp.
350 N. Diamond Street
Ravenna, OH 44266
(216) 296-9600

F, M, W

Electrical Insulation
Suppliers
1255 Collier Road NW
Atlanta, GA 30318
(404) 335-1651

A, C, F, G, I, M, P, T, W

Electro-Tech Systems, Inc.
115 E. Glenside Avenue
Glenside, PA 19038
(215) 887-2196

T, W

Engineered Static Solutions
855 Industrial Highway
Cinnaminson, NJ 08077
(609) 786-8585

F, G, I, M, P, W

EOS Protective Products
2697 FM 725
New Braunfels, TX 78130
(512) 629-7752

A, C, F, G, I, M, P, T, W

Epsco, Inc.
1265 N. Manassero
Anaheim, CA 92807
(714) 693-0114

T

Directory of Suppliers

ESD Control Co., Inc. A, C, F, G, I, M, P, T, W
13405 Floyd Circle
Dallas, TX 75243
(214) 235-1554

Espti G, M, P, W
8932 Reseda Boulevard
Northridge, CA 91324
(818) 886-0994

Exair Corporation I
1250 Century Circle N.
Cincinnati, OH 45246
(513) 671-3322

Fancort Industries, Inc. W
71 Fairfield Place
West Caldwell, NJ 07006
(201) 575-0610

Foamfab Inc. M, P
411 Oakland Street
Mansfield, MA 02048
(508) 339-2358

Foam Factory, Inc. M, P, W
2301 NW 33rd Court, No. 5
Pompano Beach, FL 33069
(305) 987-1700

Forcefield Static G, M, P, T, W
Management Systems
940 Gateway Drive
Burlington, Ontario
Canada L7L 5K7
(416) 632-6894

GC Thorsen A, G, W
1801 Morgan Street
Rockford, IL 61103
(815) 968-9661

Gemini Plastic Enterprises A
3574 Fruitland Avenue
Maywood, CA 90270
(203) 582-0901

Appendix B

Golden Star Inc. A, F, G
400 E. 10th Avenue
N. Kansas City, MO 64116
(816) 842-0233

Gordon, Packaging, L P
1050 South Paca Street
Baltimore, MD 21230
(301) 539-6537

H & S Industries A, C, F, G, I, M, P, T, W
466 E. Arrow Highway
San Dimas, CA 91773
(714) 599-6090

Hanson-Loran, Inc. A
6700 Caballero Boulevard
Buena Park, CA 90620
(714) 522-5700

Herbert Products Inc. I, T, W
180 Linden Avenue
Westbury, NY 11590
(516) 334-6500

Highland Products, Inc. P, W
River Street
Dover, NJ 07801
(201) 366-0156

Howell Packaging P
79 Pennsylvania Avenue
Elmira, NY 14902
(607) 734-6291

Industrial Custom Products M, P, W
620-622 13th Avenue S.
Hopkins, MN 55343
(800) 654-0886

Ion Company, Inc. I, T
520 Weddell Drive
Sunnyvale, CA 94089
(408) 734-2799

Directory of Suppliers

Ion Systems, Inc. I, T
2546 Tenth Street
Berkeley, CA 94710
(415) 548-3640

ITW Extruded Products M, P
Meritex Division
3301 E. Randol Mill Road
Arlington, TX 76011
(817) 649-5777

JES Ltd. A, G, I, M, P, T, W
2043 Westcliff Drive
Newport Beach, CA 92660
(714) 645-7261

Jones Associates A, C, F, G, I, M, P, T, W
228 Venture Street
San Marcos, CA 92069
(619) 471-6020

Joseph Electronics A, C, F, G, M, W
8830 N. Milwaukee Avenue
Niles, IL 60648
(312) 297-4200

Julie Associates, Inc. A, F, G, I, M, P, T, W
3 Survey Circle
North Billerica, MA 01862
(508) 667-1958

Kantek Inc. C, G, M, W
45 White Street
New York, NY 10013-2092
(212) 925-7850

Kenflex Corporation P
1034 Viking Court
Batavia, IL 60510
(312) 879-3539

Kewaunee Scientific Corp. W
P.O. Box 930
Lockhart, TX 78644-0930
(512) 398-5292

Appendix B

KeyTek Instrument Corp. T
260 Fordham Road
Wilmington, MA 01887
(508) 658-0880

Laminair Corp. C, F, G, P, W
960 E. Hazelwood Avenue
Rahway, NJ 07065
(201) 381-8200

Legge, Walter G. Co. A, F, G, M, T, W
444 Central Avenue
Peekskill, NY 10566
(914) 737-5040

Lewisystems P
Menasha Corp.
128 Hospital Drive
Watertown, WI 53094
(800) 558-9563

Line-Master Products W
14507 S. Hawthorne Boulevard
Lawndale, CA 90260
(213) 772-5255

Lissner Conductive W
Specialties Corporation
9309 Borden Avenue
Sun Valley, CA 91352
(800) 735-4500

Maine Poly Inc. A, M, P
P.O. Box 8
Greene, ME 04236
(207) 946-7440

Marshall Industries A, C, F, G, I, M, P, T, W
9674 Telstar Avenue
El Monte, CA 91731
(818) 459-5500

Melmat Inc. A, M, P, W
1400 W. 240th Street
Harbor City, CA 90710
(213) 325-1625

Directory of Suppliers

Metallized Products, Inc. M, P, W
37 East Street
Winchester, MA 01890
(617) 729-8300

Molded Fiber Glass Tray Co. P, W
East Erie Street
Linesville, PA 16424
(814) 683-4500

Monroe Electronics, Inc. T
100 Housel Avenue
Lyndonville, NY 14098
(716) 765-2254

Mustang Enterprises, Inc. P
P.O. Box 748
Geneva, IL 60134
(708) 232-1373

Nevamar Corporation M, W
8339 Telegraph Road
Odenton, MD 21113
(301) 551-5000

NRD Inc. G, I, M, P, T, W
2937 Alt Boulevard
N. Grand Island, NY 14072
(716) 773-7634

Olympic Plastics Co. M, P, W
5800 W. Jefferson Boulevard
Los Angeles, CA 90016
(213) 837-5321

Ouimet Corporation C, F, G, M, W
2967 Sidco Drive
Nashville, TN 37204
(615) 242-5478

Pandel Inc. F, W
21 River Drive
Cartersville, GA 30120
(404) 382-1034

Appendix B

Partfolio, Inc.
1281 Anderson Drive
San Rafael, CA 94901
(415) 459-1015
P, W

Patlon Industries, Inc.
11981 SW 144th Street
Miami, FL 33186
(305) 255-7744
G, M

Patlon Static Control Systems
5502 Timberlea Boulevard
Mississauga
Ontario, Canada L4W2T7
(416) 624-5572
A, C, F, G, I, M, P, T, W

Pilgrim Electric Co.
105 Newton Road
Plainview, NY 11803
(516) 420-8990
G, T, W

Plastic Systems, Inc.
261 Cedar Hill Street
Marlboro, MA 01752-9958
(508) 485-7390
A, C, F, G, I, M, P, T, W

Plug-In Storage Systems
P.O. Box 2-157
Milford, CT 06460
(203) 878-5364
A, C, F, G, I, M, P, T, W

Prevent Static Inc.
111 Wittendale Drive
Moorestown, NJ 08057
(609) 727-9699
A, G, M, P, W

Primex Plastics Corp.
65 River Drive
Garfield, NJ 07026
(201) 470-8000
M

Print Products International
8931 Brookville Road
Silver Spring, MD 20910
(800) 638-2020
A, G, T, W

Directory of Suppliers

Pro-Pac Corporation	A, C, G, I, M, P, T, W
118 Meister Avenue
Branchburg, NJ 08876
(201) 725-6696

Prudential Overall Supply	A, C, G, W
1661 Alton Parkway
Irvine, CA 92714
(714) 250-4855

Raleigh Plastics, Inc.	P, W
6312 Westgate Road
Raleigh, NC 27622
(919) 782-3603

Rawson-Lush Instrument Co.	T
80 Harris Street
Acton, MA 01720
(508) 263-3531

RDM Industrial Products Inc.	A, C, F, G, I, M, P, T, W
1829 Walsh Avenue
Santa Clara, CA 95050
(408) 727-0555

Republic Packaging Corp.	M, P
9160 S. Green Street
Chicago, IL 60620
(312) 233-6530

Richmond Technology, Inc.	A, G, M, P, W
1897 Colton Avenue
Redlands, CA 92374-9797
(714) 794-2111

Rogers Anti-Static	A
Chemicals
105 W. Madison St.
Suite 1708
Chicago, IL 60602
(312) 332-4250

Rogers Foam Corp.	A, G, M, P
20 Vernon Street
Somerville, MA 02145
(617) 623-3010

Appendix B

Santa Fe Extruders, Inc. A, M, P
15315 Marquardt Avenue
Santa Fe Springs, CA 90670
(213) 921-8991

Schaffner EMC Inc. T
825 Lehigh Avenue
Union, NJ 07083
(201) 851-0644

Scientific Enterprises Inc. A, C, G, I, M, P, T, W
P.O. Box 1295
Broomfield, CO 80038-1295
(303) 469-7801

Scientific Materials Corp. A, F, G, M, T, W
199 Garibaldi Avenue
Lodi, NJ 07644
(201) 471-1166

Seco Industries P, W
2245 East 38th Street
Los Angeles, CA 90058
(213) 589-6131

Semtronics Corporation F, G, I, M, P, T, W
P.O. Box 2248
Peachtree City, GA 30269
(800) 247-4863
P.O. Box 94
Oldwick, NJ 08858
(201) 534-5366

Sentinel Foam Products F, G, M, W
130 North Street
Hyannis, MA 02601
(508) 775-5220

Servicor Inc. M
1100 Industrial Road
San Carlos, CA 94070
(415) 591-0900

Directory of Suppliers

Set Point
31 Oxford Road
Mansfield, MA 02048
(508) 543-3800

A, M, P, W

Simco Co., Inc., The
2257 N. Penn Road
Hatfield, PA 19440-1998
(215) 822-2171

A, C, F, G, I, M, P, T, W

Spectrascan Inc.
1110A Elkton Drive
Colorado Springs, CO 80907
(719) 599-9254

G, T, W

Sportscase, Inc.
610 13th Avenue
Hopkins, MN 55343
(612) 933-4545

A, G, M, P, W

Stackbin Corp.
Albion Road
Lincoln, RI 02865
(800) 333-1603

A, W

Stanley-Vidmar Inc.
11 Grammes Road
Allentown, PA 18103
(215) 797-6600

W

Start International
15775 N. Hillcrest
Dallas, TX 75248
(214) 960-1986

A, C, F, G, I, T, W

Stat-Comp Inc.
9001 E. Bloomington Freeway
Bloomington, MN 55420
(612) 888-0271

A, C, F, G, I, M, P, T, W

Static Control Services
870 Research Drive, No. 9
Palm Springs, CA 92262
(619) 325-3211

I, T, W

195

Appendix B

Static, Inc. F, G, I, M, P, T, W
Old Sherman Turnpike
Danbury, CT 06810
(203) 791-3600

Static & Electromagnetic F, G, I, M, P, T, W
Control Division, 3M
3M Building 130-3N
P.O. Box 2963
Austin, TX 78769
(800) 328-1368

Static Prevention Inc. A, F, G, I, M, T, W
211 Gemini Avenue
Brea, CA 92621
(714) 680-6478

Static Solutions, Inc. A, F, G, M, P, T, W
258 Farrwood Drive
Bradford, MA 01835
(508) 373-4524

Stat-Tech Div. A, P, W
Gary Plastic Packaging
770 Garrison Avenue
Bronx, N.Y. 10474
(212) 893-2200

TAKK Industries Inc. I
8665 E. Miami River Road
Cincinnati, OH 45247
(513) 353-4306

Tech Spray Inc. A, G, P, W
P.O. Box 949
Amarillo, TX 79105
(806) 372-8523

Techni-Tool, Inc. A, C, F, G, I, M, P, T, W
5 Apollo Road
Plymouth Meeting, PA 19462
(215) 825-4990

Directory of Suppliers

Technibag/Dan Newman Co. A, P
1051 Bloomfield Avenue
Clifton, NJ 07012
(201) 778-6677

Terra Universal, Inc. C, G, I, M, P, T, W
700 N. Harbor Boulevard
Anaheim, CA 92805
(714) 526-0100

Trio-Tech International A, F, G, M, W
Static Systems
355 Parkside Drive
San Fernando, CA 91340
(818) 365-9200

United Foam Plastics Corp. M, P
179-F E. Main Street
Georgetown, MA 01833
(508) 352-2200

United Technical Products A, F, G, W
960 Turnpike Street
Canton, MA 02021
(800) 225-6052

V.H. Technology A, F, G, I, M, P, T, W
601 S. Jason Street
Denver, CO 80223
(303) 778-8037

Web Technologies A, M, P
27 Main Street
Oakville, CT 06779
(203) 274-9657

Wescorp Antistatic Products A, C, F, G, I, M, P, T, W
144 S. Whisman Road
Mountain View, CA 94041
(415) 969-7717

Westek F, G, I, P, T, W
400 Rolyn Place
Arcadia, CA 91007
(818) 446-4444

Appendix B

Worklon Div. C
Superior Surgical Mfg. Co.
100099 Seminole Boulevard
Seminole, FL 34648
(813) 397-9611

Z-Mar Technology A, C, F, G, I, M, P, T, W
P.O. Box 1298
Matthews, NC 28106
(704) 846-2405

References

1. DOD-HDBK 263, "Electrostatic Discharge Control Handbook for Protection of Electrical and Electronic Parts, Assemblies and Equipment (Excluding Electrically Initiated Explosive Devices)"

2. MIL-STD-883 "Test Methods and Procedures for Microelectronics Test Method 3015.6 Electrostatic Discharge Sensitivity Classification"

3. Gleeson, D. and Russeth, W., "10 Myths of Static Damage, Static and Electromagnetic Control Division/3M"

4. Gleeson, D., "Ruling Out Static in the Field Service Industry," 1986, 1987, Static and Electromagnetic Control Division/3M

5. Grasso, T. Gutman, G. and Yenni, D.M. "Triboelectric Charging Propensity of Packaging Materials," 1985, Static and Electromagnetic Control Division/3M

6. Huntsman, J.R. and Swenson D.E., "Principles of Static Protective Packaging and Work Station Design," Static and Electromagnetic Control Division/3M

7. Huntsman, J.R., and Yenni, D. M.,"Test Methods for Static Control Products," 1982, Static and Electromagnetic Control Division/3M

8. Huntsman, J.R., Yenni, D.M., and Mueller, G.E., "Fundamental Requirements For Static Protective Containers," 1980, Static and Electromagnetic Control Division/3M

9. Kunz, R.J. "The Solution To Static Caused Problems With Commercial Electronic Equipment: 3M Static Control Floor Mats," Static and Electromagnetic Control Division/3M

References

10. "Test Apparatus Used in 'Test Methods for Static Control Products', Static and Electromagnetic Control Division/3M

11. Protecting Static Sensitive Equipment in the Office Environment, Static and Electromagnetic Control Division/3M

12. KeyTek Instrument Corporation, "Electrostatic Discharge (ESD) Protection Test Handbook," Second Edition, 1986.

13. Electro-Tech Systems, Inc., product literature

14. Lacy, Edward A.,"Protecting Electronic Equipment From Electrostatic Discharge," 1984, TAB Books, Blue Ridge Summit, PA

15. Antinone, Robert J., BDM Corp. Albuquerque, N.Mex. "How To Prevent Circuit Zapping," IEEE Spectrum, April 1987, pp 34-38

16. Boxleitner, Warren, KeyTek Instrument Corp., Wilmington, MA, "How To Defeat Electrostatic Discharge," IEEE Spectrum, August 1989 pp 36-40

17. Mardiguian, Michel, "Electrostatic Discharge, Understand, Simulate and Fix ESD Problems," 1986 Interference Control Technologies Inc., Gainsville, VA

18. Vavra, Andrew L., The Simco Co., Inc., Hatfield PA, "Selecting and Using Electrostatic Fieldmeters," Evaluation Engineering, Nokomis, FL, August 1989, pp 98

19. Bhar, T.N. and McMahon, E.J.," Electrostatic Discharge Control-Successful Methods for Microelectronics Design and Manufacturing," 1983, Hayden Book Company, Inc., Rochelle Park, NJ

20. Vosteen, W. E., A Review of Current Electrostatic Measurement Techniques and Their Limitations, 1984, Monroe Electronics, Inc., Lydonville, N.Y.

Glossary of ESD Control Terminology

As in any other engineering discipline, electrostatic discharge control has its own vocabulary. Because this field combines so many different specialties, there is no agreement on all the definitions. Some of the terms defined here have not been used previously in this book, but they are included to help you understand the terminology used in references and technical papers. Many of the terms defined are used in other fields related to ESD control, such as radio frequency interference and electromagnetic interference control.

These definitions have been compiled from Electronic Industries Association Standard No. ANSI/EIA-541-1988; Military Handbook DOD-HDBK-263, *Electrostatic Discharge Control Handbook for Protection of Electrical and Electronic Parts, Assemblies and Equipment (Excluding Electrically Initiated Explosive Devices);* Electrical Overstress (EOS)/Electrostatic Discharge Symposium Proceedings; Revision of International Electrotechnical Commission (IEC) Publication 801−2; and various manufacturers' product literature.

The definitions labeled *EIA, IEC,* and *DOD-HDBK-263* have been edited slightly, but they are essentially the same as those that appear in the source documents. For exact wording, refer to those documents.

acetone   A highly flammable solvent commonly used for cleaning electronic components and equipment during manufacture. Wrist straps, shop coats, smocks, and tools should be resistant to damage from this solvent.

AES   Abbreviation for auger electron spectroscopy.

air discharge method   A method of testing, in which the charged electrode of the test generator is approached to the equipment under test (EUT), and the discharge is actuated by a spark to the EUT. (IEC)

air ionizer   A machine for supplying positive and negative air ions to nonconductive surfaces to neutralize the electrostatic charges on

Glossary

them. The use of an ionizer supplements humidification as a means for dissipating static charges.

alpha particle A positively charged particle consisting of two protons and two neutrons emitted from the nucleus of an atom during the decay process. Because of its size, it has poor penetrating ability. It cannot penetrate skin, but it can ionize air molecules. (EIA) It can also affect circuit performance. Alpha particles from ceramics have caused memory losses in semiconductor memory cells.

ampere A practical unit of electrical current. One ampere of current is 6.24×10^{18} electrons passing 1 point in 1 second. One ampere equals 1 coulomb per second (1A = 1 C/s). (EIA)

antistat, antistatic agent A chemical compound that, when impregnated in or topically applied to a primary material or substrate, renders the primary material antistatic.

antistatic materials **1.** Materials that resist triboelectric charging and produce minimal static charges when separated from themselves or other materials. *Impregnated materials* are those materials impregnated with migratory antistats. *Surface-treated materials* are those materials that have been treated by spraying, dipping, printing, or wiping with a topical antistatic agent to make them surface lubricious. **2.** ESD protective materials having a surface resistivity greater than 10^9, but not greater than 10^{14} ohms/square. (DOD-HDBK-263) **3.** A material with a volume resistivity of 10^{11} ohm-cm maximum. (MIL-STD-883) **4.** ESD-protective material having a surface resistivity greater than 10^5, but not greater than 10^{11} ohms/square. (IEC)

antistatic property The property of preventing triboelectric charge generation. A material with this property will effectively minimize the production of a static charge if it is separated from another surface. This property is not a dependent function of material resistivity or of static decay performance. (EIA)

ASTM Abbreviation for the American Society for Testing Materials.

auger electron spectroscopy A method for examining materials and surfaces in very fine detail with electron beams within a vacuum chamber.

avalanche degradation *See* thermal secondary breakdown.

bag A performed, flexible container, generally enclosed on all sides except one, that forms an opening which might be sealed after loading. It is normally constructed from one piece of material that has been folded over and sealed on two edges, or from tubing sealed at one end. (EIA)

Glossary

bipolar device A semiconductor device that employs both positive and negative carriers.

blue poly A variation of pink antistatic polyethylene.

board shunt *See* shunting bar.

body capacitance The capacitance that exists between the human body and the earth, generally between 100 and 300 picofarads. For simulation of electrostatic discharge in test Method 3015.6 of MIL-STD-883C, it is set at 100 picofarads. *See* capacitance.

body resistance The resistance of the human body, typically measured between the hands. Its value depends on several variables, including perspiration and oils. For simulation of electrostatic discharge in test method 3015.6 of MIL-STD-883C, its value is set at 1,500 ohms. *See also* resistance.

bonding The process of interconnecting conductive parts to maintain a common electrical potential. It can be done by welding, soldering, or cementing with conductive cement.

bulk breakdown A power-dependent, ESD-related failure mechanism in microelectronic and semiconductor devices and piezoelectric crystals resulting from changes in junction variables because of high-temperature spots within the junction area. These high-temperature spots cause metallization alloying or impurity diffusion, making significant changes in junction variables. One result is the formation of a resistance path across the junction. This effect is usually preceded by thermal secondary breakdown. (DOD-HDBK-263)

bulk resistivity *See* volume resistivity.

capacitance The ability of a component or material to store an electric charge. The self-capacitance of a charged conductor is the ratio of its charge to its voltage ($C = Q/V$) when all nearby conductors are grounded. The Q/V ratio depends on the voltage of nearby conductors. Capacitance is measured in units of farads. Because the farad is a very large quantity, capacitance in electronic applications is usually expressed in millionths of a farad (*microfarads*) or millionths of millionths of a farad (*picofarads*).

carrier **1.** A holder for electronic parts and devices that facilitates handling during processing, production, imprinting, or testing, and protects the parts under transport. (EIA) **2.** A basic component of topical antistats that acts as a vehicle to transport the antistatic mechanism. The carrier can either be water, alcohol, or another solvent.

catastrophic failure A sudden and complete failure, typically an open circuit or a short circuit within a device that cannot be repaired.

CDM Abbreviation for charged device model.

Glossary

charge The electrical energy stored in a capacitor or on an insulated object, measured in coulombs or fractions of a coulomb. The static charge on a body is a function of the number of separated electrons on the body (negative charge) or the number of separated electrons not on the body (positive charge). Because electrons cannot be destroyed, an electron removed from one body must go to another body, leaving behind a positive void. Thus, equal and opposite charges are always produced. (EIA)

charged device model (CDM) The model of a device that acquires a charge, which is then discharged to ground through a low-impedance path in a few nanoseconds or less. The peak current might be 10 amperes or more.

charging wand A moveable, hand-held object used to induce a charge on a surface.

clean room A separate enclosed room within a building with precisely controlled heating, air conditioning, and airborne particle filtering systems where precise processing, assembly, and inspection work is performed in an essentially dust-free environment. Clean rooms are classified according to the level of contaminants screened. These rooms are required for the processing and packaging of semiconductor devices to avoid contamination with dust particles.

CMOS Abbreviation for complementary metal-oxide semiconductor.

conductance The inverse of resistance.

conductive material 1. Materials with surface resistivities of 10^5 ohms/square or less, such as metals, bulk conductive plastics, wire impregnated materials, and conductive laminates. (DOD-HDBK-263) 2. Materials that are either surface conductive or volume conductive; the two are not necessarily interrelated. These materials are metal or impregnated with metal, carbon particles, other conductive materials, or materials whose surfaces have been treated with these substances through lacquering, plating, metallizing, or printing. A conductive material is not necessarily antistatic. A surface-conductive material will have a surface resistivity of less than 1.0×10^5 ohms/square when measured in accordance with ASTM-D-257, or equivalent. A volume-conductive material will have a volume resistivity of less than 1.0×10^4 ohm-cm when measured in accordance with ASTM-D-991, or equivalent, with the calculation based on uncompressed thickness of the material. These materials seldom charge when separated from one another, but can produce charging when separated from nonconducting surfaces. (EIA)

conductivity 1. A prime characteristic for providing protection against stationary or approaching charged bodies or people by limiting the

Glossary

accumulation of residual voltages. (DOD-HDBK-263) **2.** The ability to conduct charges.

conductor A substance or body that allows a current of electrons to pass continuously along it or through it when a voltage is applied across any two points. These materials exhibit relatively low resistivity; however, the voltage-current relationship of some materials is not always linear. (EIA)

contact discharge method A method of testing, in which the electrode of the test generator is held in contact with the equipment under test, and the discharge actuated by the discharge switch within the generator. (IEC)

corona A bluish ion discharge from a conductor, particularly with sharply pointed surfaces, that occurs when the voltage is high enough to ionize the surrounding air.

coulomb The unit of electric charge. A specific quantity of electrons (charge) on a body, expressed as 1 coulomb = 6.24×10^{18} electrons. (In electrostatics a much more practical unit is the nanocoulomb, representing a charge of 6.24×10^9 electrons.) (EIA)

coulombmeter An instrument for measuring the quantity of charge stored in a circuit or object.

Coulomb's Law One of the fundamental laws in electrostatics. The force of attraction or repulsion between two charges of electricity is proportional to the product of their magnitudes and inversely proportional to the square of the distance between them. If the charges are similar, either both positive or both negative, the force will be repulsive. If the charges are dissimilar, the force will be attractive.

coupling plane A metal sheet or plate to which discharges are applied to simulate electrostatic discharge to objects adjacent to the device under test.

current A flow of electrons past a certain point in a specified period of time, measured in amperes or fractions thereof. Current is measured in terms of electrons per second, but since this number would be tremendously large, it is usually stated in terms of coulombs per second (1 C/s = 1 A). (EIA)

current-limiting resistor A resistor inserted in an electric circuit, often as a protective device, to limit the flow of current.

decay time The time for a static charge to be reduced to a given percentage of the charge's peak voltage. It is an indirect method of measuring material surface resistivity because it is generally directly proportional to surface resistivity. (DOD-HDBK-263)

Glossary

degradation The unwanted change in the operational performance of a specimen resulting from electromagnetic interference. Degradation's not necessarily a malfunction or catastrophic failure. The ESD test specification generally requires stating the criteria for degradation of performance. (IEC)

degradation failure A gradual, partial failure of a device that occurs when one of its parameters shifts outside its specified limits.

dielectric A nonconducting material or insulator, such as air, mica, and ceramic, used to separate the plates of a capacitor.

dielectric breakdown **1.** A voltage-dependent failure mechanism in microelectronic and semiconductor devices. When a potential difference is applied across a dielectric region in excess of the region's inherent breakdown characteristics, a puncture of the dielectric occurs. It can result in either total or limited degradation of the part, depending on pulse energy. The part might heal. (DOD-HDBK-263) **2.** A threshold effect in a dielectric medium where, at some electric field strength across the medium, bound electrons become unbound and travel through the medium as a current. In solid media, the region of the current is permanently damaged. The units of measurement are usually volts per unit of thickness. (EIA)

DIP Acronym for dual-in-line package (electronic components).

DIP sticks *See* magazine.

direct application The application of the test directly to the equipment under test.

dissipative material A material with a surface resistivity equal to or greater than 1.0×10^5 ohms/square but less than 1.0×10^{12} ohms/square when measured in accordance with ASTM-D-257 or equivalent. Volume resistivity equal to or greater than 1.0×10^5 ohm-cm but less than 1.0×10^{11} ohm-cm when measured in accordance with ASTM-D-991 or equivalent with the calculation based on the uncompressed thickness of the material.

dual-in-line package (DIP) A type of package for leaded electronic devices. Semiconductor IC chips, relays, resistor networks, transformers, filters, optoelectronic devices, and other products are packaged in DIPs. The most common form of this package is a rectangular-molded plastic block with rows of pins projecting from the long sides. DIP packages are also made as premolded ceramics with rows of pins brazed along the opposing long sides of the body; they have metal or ceramic lids. Ultraviolet-erasable programmable read-only memories (EPROMs) have quartz windows in their lids to permit exposure of the chip to ultraviolet light for erasure.

DUT Abbreviation for device under test.

EAP Abbreviation for electroactive polymer.

Glossary

earth reference plane (ERP) A metal sheet or plate used as a common point for the equipment under test, the ESD generator, and the auxiliary equipment. (IEC)

edge protector *See* shunting bar.

EIA Abbreviation for Electronic Industries Association.

electrical and electronic part A part such as a microcircuit, discrete semiconductor, resistor, capacitor, thick or thin film device, or piezoelectric crystal. (DOD-HDBK-263)

electrically continuous A surface that is electrically conductive so current can be passed as the result of an applied voltage between any two points on its physical surface and when discontinuities, slots, or holes do not occupy more than 10 percent of the material's surface. (EIA)

electrically powered static eliminator An assembly that consists of two basic components: the static eliminator, generally one or more electrified needles rigidly held from a grounded metal housing or proximity ground; and the high-voltage supply, which powers the static eliminator. Ion generation occurs in the air space surrounding the highly charged needle points. (EIA)

electroactive polymer (EAP) A conductive packaging material still in the experimental stage.

electromagnetic compatibility (EMC) **1.** The ability of a device to function satisfactorily in its electromagnetic environment without introducing intolerable disturbances to that environment or to other equipment. (IEC) **2.** The ability of electronic equipment to operate without creating unacceptable electromagnetic interference or responding to such interference beyond specified limits.

electromagnetic interference (EMI) An electromagnetic disturbance caused by static sparks, lightning, radar, radio and TV transmission, motors with brush commutation, line transients, etc. EMI can induce undesirable voltage signals in electronic equipment, causing malfunction and occasionally component damage by line conduction or air propagation. Protection against EMI usually requires the use of shields, filters, and special circuit designs. *See* also shield, electromagnetic; shield, electrostatic; radio frequency interference. (EIA)

electromagnetic shield *See* shield, electromagnetic.

electrometer An instrument for measuring a potential difference or an electric charge. The instrument is typically used for static decay measurements.

electromigration A failure mechanism in fine line structures caused by the mass transport of metal by momentum exchange between thermally activated metal ions and conducting electrons. The result is

Glossary

a thinning of metal in some regions (voids) and buildup in other regions (hillocks).

electron A negatively charged elementary particle with an electrical charge equal to about 1.6×10^{19} coulomb. (EIA)

electroscope A simple test instrument used to detect an electric charge, and, when calibrated, to measure potential difference. It usually consists of two thin narrow strips of metal, such as gold leaf or aluminum foil, attached to and suspended from a common conductive shaft insulated from a metal-covered glass container. When charged, the ends of the strips will spread apart. The separation between the strips is proportional to the charge.

electrostatic charge An electric charge on the surface of an insulated object.

electrostatic detector An instrument used to determine the presence or absence, polarity, and relative magnitude of electrostatic charges in the work area. Electrostatic detectors include: electrometer amplifiers, electrostatic voltmeters, electrostatic fieldmeters, and gold leaf electroscopes.

electrostatic discharge (ESD) **1.** A transfer of an electrostatic charge between bodies at different electrostatic potentials caused by direct contact or induced by an electrostatic field. (DOD-HDBK-263) **2.** A rapid transient.

electrostatic discharge sensitivity (ESDS) The relative tendency of a device's performance to be degraded by ESD. (MIL-STD-883)

electrostatic discharge simulator Equipment used to simulate the discharge of static electricity from the human body; the naturally occurring discharge or variations of that discharge.

electrostatic field **1.** The region surrounding an electrically charged object in which another electrical charge can be induced and will experience a force. Quantitatively, it is the voltage gradient between two points at different potentials. (EIA) **2.** A voltage gradient between an electrostatically charged surface and another surface of a different electrostatic potential. (DOD-HDBK-263)

electrostatic fieldmeter An instrument used to measure the electrostatic field produced by a charged body using a noncontact probe or sensor, and to provide electrostatic field strength or electrostatic voltage readings at a calibrated distance from a charged body. (DOD-HDBK-263)

electrostatic overstress simulator *See* electrostatic discharge simulator.

electrostatics The branch of physics that deals with electricity at rest.

electrostatic shield *See* shield electrostatic.

electrostatic shielding **1.** Materials that are capable of attenuating or shunting an electrostatic field so that its effects do not reach the

Glossary

stored or contained item and produce damage. *See* Faraday cage; shield, electrostatic. (EIA) **2.** Typically, nickel- or aluminum-coated polyethylene packaging material that gives Faraday cage protection from external static fields and discharges.

EMC   Abbreviation for electromagnetic compatibility.

EMI   Abbreviation for electromagnetic interference.

energy   The ability to do or perform work. Electrical energy is measured in watt-seconds, joules, or fractions thereof. A spark is energy being pended. Because a joule is a large quantity of energy and because electric sparks do not usually have energy measurable in joules, their energy is usually measured in thousandths of a joule, *millijoule*. A static spark in the order of millionths of a joule, *microjoule*, will damage semiconductors. Electrostatic energy is calculated from the relations of $CV^{2/2}$, $Q^{2/2}C$, or $QV/2$. (EIA)

energy storage capacitor   The capacitor of the ESD generator representing the capacity of a human body charged to the test voltage value. This might be provided as a discrete component, or as a distributed capacitance. (IEC)

EOS   Abbreviation for electrical overstress.

equipotential bonding   Bonding that causes all conductors to have the same potential.

ERP   Abbreviation for earth reference plane.

ESD   Abbreviation for electrostatic discharge.

ESD-protective material   Material capable of one or more of the following: limiting the generation of static electricity, rapidly dissipating electrostatic charges over its surface or volume, or providing shielding from ESD spark discharge or electrostatic fields. ESD-protective materials are classified in accordance with their surface resistivity (or alternate conductivity) as conductive, static-dissipative, or antistatic. (DOD-HDBK-263)

ESD-protective packaging   Packaging made with ESD-protective materials to prevent ESD damage to ESDS items. (DOD-HDBK-263)

ESDS   Abbreviation for electrostatic discharge sensitivity.

ESD-sensitive device   **1.** All microelectronic devices are susceptible to ESD to some extent, but for the purpose of the MIL-STD-883 test method and the associated requirements, an ESDS device is one which can be damaged by exposure to ESD at a level less than 10,000 volts using the standard circuits shown in the standard. (MIL-STD-883) **2.** Electrical and electronic parts, assemblies, and equipment that are sensitive to ESD voltages of 15,000 volts or less as determined by the test circuit given in DOD-HDBK-263. (DOD-HDBK-263)

EUT   Abbreviation for equipment under test.

Glossary

Faraday cage An electrically continuous, conductive enclosure that provides electrostatic shielding for a region with no electrostatic field. The cage or shield is usually grounded, although it need not be. (EIA)

finger cots A latex or similar covering for fingers that prevents contamination of materials being handled in clean room conditions. Conventional latex finger cots can generate static charges when rubbed against an object.

gaseous arc discharge A voltage-dependent ESD-related failure mechanism in microelectronic and semiconductor devices. For parts with closely spaced, unpassivated thin electrodes, gaseous arc discharge can cause degraded performance. The arc discharge condition causes vaporization and metal movement that is generally away from the space between the electrodes. (DOD-HDBK-263)

gauntlet A protective sleeve of static-dissipative material with an elasticized cuff at one end. It extends from the bare wrist to the elbow and is used to cover long-sleeved apparel not made of a static-dissipative fabric.

GFI Abbreviation for ground-fault interrupter.

ground **1.** A metallic connection with the earth to establish zero potential or voltage with respect to ground or earth. It is the voltage reference point in a circuit. It is understood that a point in the circuit said to be at ground potential could be connected to earth without disturbing the operation of the circuit in any way. Grounds that can be used for static-control workstations include water pipes, any power ground, or any large metal structural member of a building, vessel hull, etc. (EIA) **2.** A mass such as earth or a ship hull or vehicle capable of supplying or accepting a large electrical charge. (DOD-HDBK-263)

grounded workbench A work surface of a table or bench in contact with ESD-sensitive items made of static-dissipative materials over the area where the sensitive items would be placed. This surface should be connected to ground through a ground cable. (DOD-HDBK-263)

ground-fault interrupter (GFI) A circuit or device that senses leakage current from faulty test equipment and interrupts the circuit almost instantaneously when these currents reach a potentially hazardous level. (DOD-HDBK-263) *Also called* ground-fault circuit interrupter.

ground, hard A connection to ground, either directly or through a low impedance. (DOD-HDBK-263)

grounding The act of connecting to ground or to a conductor that is grounded; a means of referencing all conductive objects to a zero-

Glossary

voltage equipotential surface. This is the surest method of eliminating ESD since everything is maintained at the same potential. (EIA)

grounding strap   *See* personnel ground strap.

ground, soft   A connection to ground through an impedance sufficiently high to limit current flow to safe levels for personnel—normally 5 milliamperes. Impedance needed for a soft ground is dependent on the voltage levels that could be contacted by personnel near the ground. (DOD-HDBK-263)

handled or handling   The action of grasping items by hand for manipulation or for machine processing during work such as inspection, manufacture, assembly, processing, testing, repairing, reworking, maintaining, installing, transporting, failure analyzing, wrapping, packaging, marking, or labeling. (DOD-HDBK-263)

HBM   Abbreviation for human-body model.

HCMOS   Abbreviation for high-speed silicon-gate CMOS.

healing   The self-correction of a fault in electrical characteristics in damaged semiconductors and capacitors, typically damage to a dielectric.

high-voltage relay   A relay built to withstand high voltages on its contacts and used in many ESD simulators.

HMOS   Abbreviation for high-density MOS.

holding time   The time of the decrease of the test voltage resulting from leakage, prior to the discharge; not to be greater than 10 percent.

human capacitance   *See* body capacitance.

human resistance   *See* body resistance.

humidity control   The process of adjusting the relative humidity in the work environment to minimize electrostatic charge. In general, the higher the relative humidity (RH), the lower will be the ESD problem to about 60 percent RH. However, high RH can make working conditions intolerable for personnel and accelerate oxidation of tools and materials.

humidity test chamber   An enclosure with controlled humidity used for the performance of static-decay tests.

hygrometer   An instrument for measuring atmospheric humidity.

IC   Abbreviation for integrated circuit.

IFM   Abbreviation for ion flux monitor.

immobile charge   The charge residing on nonconductors.

immunity   The ability of a device, equipment, or system to resist an electromagnetic disturbance.

Glossary

indirect application The application of a test to a coupling plane in the vicinity of the equipment under test that simulates personnel discharge to objects adjacent to the EUT. (IEC)

induction The process of establishing an electric charge on a nearby object without physical contact.

induction static eliminator An apparatus consisting of a series of conductive grounded points or brushes. When a single grounded needle point is brought into the proximity of a highly charged surface, it has induced in it a charge opposite to that of the surface. When a high charge concentration has been developed, the surrounding air will break down, forming a vast number of charge-balancing ions. The simple tinsel static eliminator is an example of an induction static eliminator. (EIA)

input protection network Resistors, capacitors, diodes, transistors, or any combination of these devices placed at the input of an IC to protect it from electrostatic discharge.

insulative material 1. A material having a surface resistivity of 10^{14} ohms/square minimum, or a volume resistivity of 10^{12} ohm-cm minimum. (MIL-STD-883) 2. A material with a surface resistivity equal to or greater than 1.0×10^{12} ohms/square or a volume resistivity equal to or greater than 1.0×10^{11} ohm-cm.

insulator A material that is a poor conductor or nonconductor of electricity. (EIA)

integrated circuit (IC) Any monolithic semiconductor device containing transistors, resistors, capacitors, and other components on or in a single substrate or chip. To define an IC further, modifiers are used. For example, a bipolar IC is one containing junction semiconductors that use both positive, *p-type*, and negative, *n-type*, charge carriers. A MOS or CMOS IC is one whose transistors depend on a single-charge carrier, p-type or n-type.

intimate packaging material Materials in direct contact with the item packaged.

ion flux monitor (IFM) An instrument for determining the amount of air ions available from an ionizing source.

ionization The process by which a neutral atom or molecule, such as air, acquires a positive or negative charge. (EIA)

isopropanol A solvent commonly used in the electronics industry. Bench or table surfaces should be resistant to damage from this chemical.

joule A unit of energy. The energy of a statically charged conductive object is $1/2\ CV^2$ joule, where C = capacitance of the object and V = its voltage. (EIA)

Glossary

latent failure A failure that occurs in a device some time after it has been exposed to ESD. At the time of the discharge, there was no apparent damage.

lubricity A measure of surface smoothness and lubricating action of moistness. Triboelectric generation is a friction process; the higher the lubricity of the surfaces being rubbed, the lower the friction; hence the lower the generated charges. (DOD-HDBK-263)

magazine A rigid extruded package with a C-shaped cross section for the shipping and storing of DIP-packaged integrated circuits. *Also called* DIP stick, slide, or rail.

metallization melt A power-dependent ESD-related failure mechanism in microelectronic and semiconductor devices occurring when ESD transients increase part temperature sufficiently to melt metal or fuse bond wire. (DOD-HDBK-263)

metalloplastics Plastics that have metallic fibers added. The composite has a low resistivity, thereby reducing the inherent problem of static charges forming on plastics.

MNOS Abbreviation for metal-nitride-oxide semiconductor.

mobile charge The charge resting on the conductive parts of an object.

MOS Acronym for metal-oxide semiconductor.

MOS device structure **1.** A device with a metal-oxide semiconductor layer structure. The oxide is a thin dielectric that is voltage sensitive and can be easily damaged by a static discharge or by induction from an electrostatic field. The damage occurs as a dielectric breakdown when the voltage or electric field across the dielectric layer between two relatively conductive layers exceeds the dielectric strength of that material. (EIA) **2.** A conductor and a semiconductor substrate separated by a thin dielectric. (DOD-HDBK-263)

MOSFET Acronym for metal-oxide semiconductor field-effect transistor.

MOV Abbreviation for metal-oxide varistor.

nanocoulomb One-billionth (10^{-9}) of a coulomb. *See* coulomb.

NMOS Abbreviation for N-channel MOS.

nonconductor *See* insulator.

nonintimate packaging materials Packaging materials that do not come in direct contact with the device or assembly being packaged, or are outside the intimate packaging material. (EIA)

nuclear ionizer *See* nuclear static eliminator.

nuclear static eliminator An apparatus used to create ions by the irradiation of the air molecules. Some models use a safe alpha-emitting isotope to create sufficient ion pairs to neutralize a charged sur-

Glossary

face. The high-speed particle interacts with air molecules with sufficient energy to actually strip off one of its outer electrons. (EIA) *See* ionization.

ohm The unit of electrical resistance. It is the resistance through which a current of one ampere will flow when a voltage of one volt is applied. (EIA)

ohms per square The unit of surface resistivity. For any material, it is numerically equal to the surface resistance between two electrodes forming opposite sides of a square, regardless of its size. It is normally used as a resistivity measurement of a thin conductive layer or material over a relatively insulative base material. (DOD-HDBK-263)

ozone-triatomic oxygen An unstable, colorless, or pale blue gas with a pungent characteristic odor. It can deteriorate rubber insulation on electrical conductors. At concentrations higher than 0.1 part per million, it can be hazardous to the health. It is produced as an undesired by-product during the ionization of air by high-voltage ionizers. The amount produced by commercial ionizers is generally at a low enough level to present no hazard for nearby persons.

packaging The enclosure of products, devices, or other packages in a wrap, pouch, bag, slide, magazine, or other container to perform one or more of the following functions: a. Containment for storage, handling, and transportation. b. Preservation and protection of the contents for the life of the item. c. Identification of contents including quantity and manufacturer. d. Facilitate the dispensing and use of the contents. (EIA)

packaging materials Those materials that cushion, enclose, or protect the finished product during transportation and storage including bags, boxes, wraps, cushioning materials, foams, and magazines (slides, tubes, rails). (EIA)

packing density In an integrated circuit, the number of gates per unit area on a semiconductor substrate or chip.

PCB Abbreviation for printed-circuit board.

personnel apparel Recommended apparel, including long-sleeved protective smocks or close-fitting, short-sleeved shirts or blouses, for people handling ESD-sensitive items. (DOD-HDBK-263)

personnel ground strap A skin-contact wrist, leg, or ankle ground strap to be worn by persons handling ESD-sensitive items. Their purpose is to dissipate rapidly and safely to ground all static charges from the wearer, and to equalize personnel static levels with that of the work surface. (DOD-HDBK-263) *See also* wrist strap.

pink poly Abbreviated form of pink polyethylene, a plastic-impregnated antistat.

Glossary

plastic bubble wrap/pack A type of two-ply protective wrap or bag-packaging material made by selectively sealing small pockets or cells of inert gas between the two plies. The cells act as shock absorbers and thermal insulators. The opposing quilted surfaces of these bags makes them easier to open than those made of smooth-ply materials.

PMOS Abbreviation for P-channel metal-oxide semiconductor.

potential Voltage, measured in millivolts, volts, or kilovolts, from a base point. This point can be any voltage, but is usually ground—a theoretically zero voltage. (EIA)

pouch A small or moderately sized plastic bag made by sealing three edges of two flat sheets of flexible material, or by sealing one end of a tube of flexible material. (EIA) *See also* bag.

protected area An area constructed and equipped with the necessary ESD-protective materials, surfaces, and equipment to limit ESD voltage under the sensitivity level of ESD-sensitive items handled there. (DOD-HDBK-263)

protective flooring Commercially available flooring in the form of conductive, static-dissipative, and antistatic mats, carpeting, sheeting, and tiles.

protective handling Handling of ESD-sensitive items in a manner to prevent damage or destruction from ESD.

PVC Abbreviation for polyvinyl chloride.

radioactive ionizers *See* nuclear static eliminator.

radio frequency interference (RFI) A part or subset of electromagnetic interference. (EIA) *See* electromagnetic interference.

resistance The measurement of the difficulty an electrical charge encounters in passing from one point to another. Resistance is a property of dimensions, surface and volume resistivity, temperature, and also voltage in nonohmic materials. The resistance of the material determines the current (electron flow) a given voltage produces. The practical unit of resistance is the ohm. (EIA)

resistivity A measure of the intrinsic ability of a material to conduct current. Its value is independent on the dimensions of the material. Both conductors and nonconductors have resistivity. The unit of volume resistivity is the ohm-cm. The unit of surface resistivity is ohms/square. (EIA)

resistivity, surface 1. The ratio of dc voltage to the current that passes across the surface of the system. In this case, the surface consists of a square unit of area. In effect, the surface resistivity is the resistance between the two opposite sides of a square and is independent of the size of the square or its dimensional units. Surface resistivity is expressed as ohms/square. When used with a concen-

Glossary

tric ring fixture, resistivity is calculated by using the following expression:

$$\text{surface resistivity: } (\varrho_s) = \frac{\pi (D_2 + D_1)}{(D_2 - D_1)} \times R$$

where: D_2 = Inside diameter of outer electrode
D_1 = Outside diameter of inner electrode
R = Measured resistance in ohms   (EIA)

2. An inverse measure of the conductivity of a material and equal to the ratio of the potential gradient to the current per unit width of the surface, where the potential gradient is measured in the direction of current flow in the material. For any material, it is numerically equal to the surface resistance between two electrodes forming opposite sides of a square, regardless of the size. Surface resistivity applies to both surface- and volume-conductive materials and has the value of ohms/square. It normally is used as a resistivity measurement of a thin conductive layer of material over a relatively insulative base material. It is not constant for a homogeneous material, but varies with material thickness. Therefore, the relationship of surface resistivity to volume resistivity is meaningless for a homogeneous bulk conductive material, unless the thickness is also given. It is used to measure the resistivity of surface-conductive materials such as hygroscopic antistatic polyethylene, nylon, virgin cotton, metal- or carbon-coated paper, plastics, and other conductively coated or laminated insulative materials. (DOD-HDBK-263)

resistivity, volume **1.** The ratio of the dc voltage per unit of thickness, applied across two electrodes in contact with a specimen, to the amount of current per unit area passing through the system. Volume resistivity is generally given in ohm-cm. When using a concentric ring fixture, resistivity is calculated by using the following expression:

$$\text{Volume Resistivity: } (\varrho_v) = \frac{\pi D_1^2}{4T} \times R$$

where: D_1 = Diameter of inner electrode or disk
R = Measured resistance in ohms
T = Thickness of specimen   (EIA)

2. An inverse measure of the conductivity of a material equal to the ratio of the potential gradient to the current density, where the potential gradient is measured in the direction of current flow in the material. (Note that in the metric system, volume resistivity of an electrical insulating material in ohm-cm is numerically equal to the volume resistance in ohms

Glossary

between opposite faces of a 1-cm cube of the material.) It is a constant for a given homogeneous material. *Also called* bulk resistivity. (DOD-HDBK-263)

SAW   Abbreviation for surface acoustic wave device.

SDP   Abbreviation for static discharge pulse.

SEM   Abbreviation for scanning electron microscope.

sensitive device   *See* ESD-sensitive device.

sensitivity   The minimum value that a sensor will detect effectively and reliably. (EIA)

sheet resistivity   *See* surface resistivity.

shield, electromagnetic   A screen or other housing placed around devices or circuits to reduce the effects of both electric and magnetic fields. The electromagnetic field results from the presence of a rapidly changing electric field and its associated magnetic field. Shielding from electromagnetic interference is achieved with a combination of reflection and absorption of electromagnetic energy by the material. Reflection occurs at the surface much like the reflection of light at an air-to-water interface, and is not usually affected by shield thickness. Absorption, however, occurs within the shield and is highly dependent on thickness. (EIA) An electromagnetic should not be confused with an electrostatic shield.

shield, electrostatic   A barrier or enclosure that prevents the penetration of an electrostatic field. An electrostatic shield might not offer much protection against the effects of electromagnetic interference (EMI). EMI shields, however, are good electrostatic shields. *Compare* shield, electromagnetic.

shielding   *See* shield, electromagnetic; shield, electrostatic.

shipping tubes   *See* magazine.

shuffle test   A test to evaluate the static-generation properties of flooring materials involving sliding a foot across the floor.

shunting bar   Apparatus for shorting together the terminations of conductive strips on both sides of a circuit board. The terminations are brought in close proximity for coupling to external circuitry with card-edge connectors. Shunting bars are also called *edge protectors* or *board shunts*.

smock   A garment worn over street clothes to dissipate static charges on clothing.

SOS   Abbreviation for silicon-on-sapphire.

spark   An abrupt, short-duration electric discharge that causes a visible flash of light. The electromagnetic pulse caused by ESD discharge in the form of a spark can cause part failure and upset digital equipment, such as computers. (DOD-HDBK-263)

Glossary

Speidel band A type of wrist band with spring-loaded links named for the expandable metal watch bands made by the Speidel Company. The bands provide dependable contact with the user's skin, making an effective body ground, which is important in ESD control. The outer surfaces of the links are coated with an insulating material to prevent shock if the user contacts a live electrical circuit.

static awareness A desired attitude for people who work with ESD-sensitive devices and components, whether in manufacture, assembly, test, or repair. Because they are convinced that a threat to products exists, they follow procedures and wear special clothing when handling ESD-sensitive devices.

static-dissipative materials ESD-protective material with surface resistivity greater than 10^5 but not greater than 10^9 ohms/square. Static-dissipative materials could include conductive materials, except that the thicknesses are lower; wire or wire mesh included therein is finer, or there is more space in conductive materials; or volume resistivities are higher. (DOD-HDBK-263)

static electricity Electricity that is not moving.

static electricity discharge A transfer of electrostatic charge between bodies of different electrostatic potential.

static eliminator, electrical One or more electrified needles rigidly held less than 1 inch from a grounded metal housing or proximity ground, together with the high-voltage supply that powers the static eliminator. Ion generation from electrical static eliminators occurs in the air space surrounding the highly charged needle points.

static eliminator, induction A series of conductive-grounded points or brushes. When a single grounded needle point is brought close to any highly charged surface, it has induced in it a charge opposite to that of the surface. When a high enough charge concentration has been developed, the surrounding air will break down, forming a vast number of charge-balancing ions. The simple *tinsel static eliminator* is an example of an induction static eliminator.

static eliminator, nuclear An eliminator that creates ions by the irradiation of the air molecules. Some models use an alpha particle-emitting isotope to create sufficient ion pairs to neutralize a charged surface. The high-speed particle interacts with air molecules with sufficient energy to strip off one of its outer electrons. (EIA) *See* ionization.

static shielding *See* electrostatic shielding.

surface breakdown A voltage-dependent, ESD-related failure mechanism in microelectronic and semiconductor devices. For perpendicular junctions it is a localized avalanche multiplication process caused by a narrowing of the junction space charge layer at the surface. (DOD-HDBK-263)

surface resistivity *See* resistivity, surface.

Glossary

thermal secondary breakdown A power-dependent, ESD-related failure mechanism in microelectronic and semiconductor devices, also known as *avalanche degradation*. Since thermal time constants of semiconductor materials are generally large compared with transient times associated with ESD pulses, there is little diffusion of heat from the areas of power dissipation, so large temperature gradients can form in the parts. Localized junction temperatures can approach material melting temperatures, usually resulting in the development of hot spots and subsequent junction shorts because of melting. (DOD-HDBK-263)

transient voltage suppressor (TVS) A device that can reduce the voltage and energy flowing into an electrical circuit to levels sufficient to avoid damage to parts in the assembly levels. Suppressors include tin-, zinc-, or bismuth-oxide, voltage-dependent resistors, *VDRs*. TVSs are also known as referred to as metal-oxide varistors, *MOVs*, and silicon voltage limiters or suppressors, also known as *transient voltage suppressors*. (DOD-HDBK-263)

triboelectric charge An electrical charge generated by frictional movements or separation of two surfaces. (EIA)

triboelectric series **1.** A list of substances in an order of positive to negative charging as a result of the triboelectric effect. (DOD-HDBK-263) **2.** A list of substances arranged so that one can become positively charged when separated from one farther down the list, or negatively charged when separated from one farther up the list. The series main utility is to indicate likely resultant charge polarities after triboelectric generation. However, this series is derived from specially prepared and cleaned materials tested in controlled conditions. In everyday circumstances, materials reasonably close to one another in the series can produce charge polarities opposite to that expected. (Note: This series is only a guide.) (EIA)

trichloroethane A common solvent used in the electronics industry. Wrist straps, trays, work surfaces, and other materials used in an ESD-controlled area should be able to withstand contact with this solvent.

TVS Abbreviation for transient voltage suppressor.

uniforms In ESD control, commercially cleaned static-controlled garments provided by the company, particularly used in rooms that are also certified clean rooms.

upset failure An intermittent failure in the operation of a semiconductor device caused by ESD, that could make it necessary to shut off the power and restart the circuit.

V Symbol for volt.

Glossary

Van de Graaff generator A high-voltage electrostatic generator for physics research in lightning simulation and as a high-energy source for nuclear experiments.

VDR Abbreviation for voltage-dependent resistor.

VLSI Abbreviation for very-large-scale integration (integrated circuit).

VMOS Abbreviation for vertical-groove MOS, a power transistor structure.

volt (V) The unit of voltage, potential, and electromotive force. One volt will send a current of one ampere through a resistance of one ohm. (EIA)

voltage The electrical potential difference that exists between two points and is capable of producing a flow of current when a closed circuit is connected between the two points. (EIA)

voltage-dependent resistor (VDR) A transient suppressor, also referred to as a metal-oxide varistor. (DOD-HDBK-263)

voltage suppression A phenomenon in which the voltage from a charged object is reduced by increasing the capacitance of the object, rather than decreasing the charge on the object. The relation $Q = CV$ describes the phenomenon. It occurs most frequently when a charged object is close to a ground plane, but is not in resistive contact with the ground plane. (EIA)

volume resistivity *See* resistivity, volume.

wound A degradation, in contrast to hard failure, of semi-conductor devices.

wrist strap A close-fitting, conductive cuff circling the wrist in the fashion of a watch band that is connected to ground through a current-limiting resistor. By conducting body charges to ground, it provides a first line of defense against ESD damage.

Index

A

A/B comparisons, 9
ac-carrier fieldmeter, 129
air guns, ionizing, 109
alpha particles, 108
amber, 16
antenna effect, 149
antistatic material development, 11
antistatic materials (*see also* containers and packaging), 11, 22, 25, 26, 55, 78, 95, 96
   classification of, 78-83, 95, 96
   development of, 11
   topically applied, 101-103, 111-112
arcing, 6, 66, 76
atomic theory, 16-17, 108
attraction, electrostatics, 17-19, 24, 32, 37-40, 104
avalanche degradation, 61

B

bench-type simulators, 154
bipolar transistors, 6, 7, 57, 61, 62, 69
bound charges, 48
breakdown voltages, 69
burnout, 6, 63-64, 67

C

cables and wires, shielding, 76
capacitance/capacitors, 44-50, 69, 75
   bound charges, 48
   capacitor types, 48-49
   charge storage, 44
   determination of, 45
   dielectrics in, 44-48
   energy storage in, 45
   grounding vs., 50
   parallel-plate, 46
   parasitic capacitors, 49-50
   polarization, 47
   unintentional capacitors, 49-50
capacitative coupling, 7, 66
card edge protectors, 91-92
chair mat, ESD-protective, 102, 103
charges, 15-16
   bound, 48
   dissipation of, 26-29, 55
   equipotential surfaces, 41-43
   induction-caused, 29-30, 39
   mobile vs. immobile, 22, 30
   point, 19
   positive vs. negative, 17-19, 24, 32, 37-40, 104
   storage of, 44
   surface charge density, 43-44
   triboelectric, 20-24, 77, 79, 126
Charleswater Products, 108
chopper-stabilized fieldmeter, 129
circuit board shunts, 91-92
clothing, 9, 121-124, 161-166
CMOS, 2, 8, 57, 63, 66, 69, 70, 71, 72, 150
coatings/finishes, ESD-protective, 101-103, 111-112

Index

conductive coupling, 76
conductive materials (*see also*
    containers and packaging), 25,
    78-83, 95, 96
conductive shunts, 77
conductive (protective) work surfaces,
    1, 9-10, 27, 68, 83,
    95-104, 159, 161-166
  chair mat, 102, 103
  electric shock hazards vs., 98-99
  floors and finishes, 101-103
  ground potential of, 100
  ground-fault interrupters (GFI), 99
  mechanical considerations, 100
  seating/cushions, 104
  soft grounds, 95
  table mat, 98
  test equipment/tools, grounding, 99
  workbenches, 95-98
conductors, 6, 22, 24-28, 30, 57
connectors, 7
containers and packaging, 1, 4, 10,
    27, 30, 31,
    68, 76-83, 159, 166, 167
  antistatic material, 78-83
  card edge protectors, 91-92
  circuit board shunts, 91-92
  conductive materials, 78-83
  DIP magazines/DIP sticks, 84-88
  Faraday cage, 83-85
  materials for, classification of, 78-83
  protective bags, 83-85
  protective foams, 90-91
  shunt bars, 91-92
  static-dissipative material, 78-83
  tote boxes/storage cases, 88-90
control program, 159-175
  audit provisions for, 172-173
  conductive work surfaces, 161-166
  customer responsibilities, 175
  general handling procedures and
    requirements, 166
  inspection procedures, 170-171
  preparation and monitoring for,
    160-161
  printed circuit boards, handling
    guidelines, 169-170
  product design, 174-175
  semiconductor devices, handling
    guidelines, 167-169
  specifications for, 162-164
  testing, field, 171

training and certifying personnel,
    173-174
copying machines, 11
corona discharge, 44, 55, 151
cost of repairs from ESD, 2-3, 8
Coulomb's Law, 19, 32
Coulomb, Charles, 19
coupling, 75
  capacitative, 7, 66
  conductive, 76
  inductive, 7, 66
  radiated, 6
current-limiting devices, 69
customer responsibility, ESD
    protection, 175

D

damage detection, 2, 6, 8-10, 60-63
  electron microscope analysis, 64-65
  erroneous signals, 66
decay time, 27, 126
Department of Defense (DOD), 11,
    27, 50, 56, 78, 95, 96, 150, 152,
    153
dielectric breakdown, 6, 62-63
dielectrics, 44-48
DIP sticks/DIP magazines, 30-31,
    84-88
dissipation of ESD charge, 26-29, 55
DOD-HDBK-263, 57, 78, 79, 95, 96
dust accumulation, ESD and, 10

E

E-fields, 151
EIA Standard RS-541, 78, 79, 89
electric fields, 30-35, 77, 81, 126
electric potential, 40, 41, 44
electric shock hazards, 98-99,
    114-125
  ground fault interrupters (GFI), 99
  hot-bar ionizers, 106-108
  parallel paths, 99
electric-powered ionizers, 105
electricity, 10
Electro-Tech Systems, 143, 146, 147,
    154
electrolytic capacitors, 48
electromagnetic interference (EMI),
    5, 12, 13, 175
electromagnetic radiation, 65
electrometer, 129

222

Index

Electronic Industries Association (EIA), 27, 78
electrons, 16-17
electroscope, 34-40
electrostatic discharge (ESD)
  antistatic material development, 11, 22
  benefits of, 11
  components of charges, 151
  containers and packaging vs., 77-83
  corona discharge, 44, 55
  cost of repairs from, 2-3, 8
  damage detection, 2, 6, 8-10, 60, 61-66
  defined, 5-7, 10-11
  discovery of, 16
  duration of transients, 13, 14, 55, 151
  electrostatic principles defined, 15-55
  environmental factors contributing to, 11-14, 68, 93-113
  filtering, 75, 76
  frequencies of, 13
  gate density vs., 2-3
  hand/metal, 151
  levels of, 1, 4-10, 15, 20-24, 28, 50
  lightning as, 10, 12, 13, 14, 44, 66
  lubricity vs., 22, 79, 81
  personal protection against, 114-125
  precautions against, 1, 4-5, 9-10, 27, 55, 68
  product design vs., 174-175
  protective devices, 4, 69-76, 149, 150
  relative humidity vs., 20-24, 26, 28, 55, 68, 79, 81, 82, 93-94, 145-146
  semiconductor design vs., 2-5, 7-9, 11, 66, 69-76, 149, 150
  shielding, 76
  software damage by, 76
  sources of, 1, 10, 12, 13, 14, 20-24
  standards of testing and measurement, 152-153
  static awareness, 10
  static electricity, history of, 10-11
  storage of, human body and, 6, 28, 50, 51, 52, 150
  test equipment for, 126-148
  transmission of, 6, 23-24, 28, 30, 50-52, 150
  universality of problem, 8-10
electrostatics, 15-55
  antistats vs., 22
  atomic structure and, 16-17
  attraction and repulsion, 17-19, 24, 32, 37-40, 104
  capacitance/capacitors, 44-50
  charge dissipation, 26-29, 55
  conductors and insulators, 24-26
  Coulomb's Law, 19, 32
  decay time, 27, 126
  electric potential, 40, 41, 44
  equipotential surfaces, 41-43
  Faraday cage, 32, 33, 34, 83-85
  fields of electricity, 30-35, 77, 81, 126
  grounding, 28, 41, 95
  hard grounds, 29
  human body ESD simulator, 50-55
  induction charging, 29-30, 39
  lubricity vs., 22, 79, 81
  measurement of, electroscope for, 34-40
  point charges, 19
  polarity, 39
  potential difference, 40, 41, 44
  resistance, 24, 25, 78
  sheet resistance, 25
  soft grounds, 28, 29, 95
  surface charge density, 43-44
  surface resistivity, 25, 26, 78, 79, 142, 143
  triboelectric charging, 20-24, 28, 77, 79, 126
  Van de Graaf generator for, 50-52
  voltage, 41, 44
  volume resistivity, 24, 25, 78, 142, 143
emitter-coupled logic (ECL), 7
enclosures, 76
environmental factors, ESD and (see also relative humidity), 11-14, 68, 93-113
  auditing conditions, 172-173
  clothing, 161, 166
  conductive work surfaces, 95-104
  gloves and finger cots, 161
  humidity control, 93-94
  ionizers, 94, 104-109
  tools/production equipment, 110-111
  topical antistats, 111-112

223

Index

environmental factor (*cont.*)
    warning labels/signs, 113, 166, 167
    wrist straps, 161
equipotential surfaces, 41-43
erroneous signals, 66
European Computer Manufacturer's Association (ECMA), 153

F

Faraday cage, 32, 33, 34, 83-85
Faraday, Michael, 32, 45
field service kits, ESD-protective, 124-125
fieldmeters, 126-140
fields, electric, 30-35, 77, 81, 126, 151
fields, magnetic, 152
filters, 4, 75, 76
finishes/coatings, ESD-protective, 101-103, 111-112
floors, ESD-protective, 101-103
foams, ESD-protective, 90-91
Franklin, Benjamin, 44
frequencies, ESD, 13
friction (*see also* lubricity), 10, 15, 20-24, 79

G

gallium arsenide (GaAs) transistors, 2, 57, 66, 150, 167-169
gate density, 2-3, 7, 9, 70, 149
Gilbert, William, 16
gloves and finger cots, ESD-protective, 124, 161
ground strap testers, 126, 147-148
ground-fault interrupters (GFI), 99
grounding, 4, 28, 41, 55, 68, 83, 95
    capacitance vs., 50
    electric shock hazards and, 98-99
    ground fault interrupters (GFI), 99
    ground potential, 100
    hard ground, 29
    parallel paths vs., 99
    resistance to, 126
    shoe straps for, 120-121, 147
    soft grounds, 28, 29, 95
    tools/test and production equipment, 99, 110-111
    wrist straps for, 114-120, 147

H

H-fields, 152
hand-held simulators, 155-157
hand/metal ESDs, 151
hard failure, 61-64
    avalanche degradation, 61
    dielectric breakdown, 62-63
    junction burnout, 61
    metallization melt, 63-64, 67
    oxide punchthrough, 62-63, 69
    thermal secondary breakdown, 61-62
hard grounds, 29
hot/cold bars, ionizers, 106-107
Hughes Aircraft Corp., 11
human body ESD simulator, 50-55, 150, 152, 153
humidity (*see* relative humidity)
humidity test chambers, 126, 145-146
hybrid microcircuits, 57
hygroscopic chemicals, 26, 81

I

IEC 801-2, 79, 152
immobile charges, 22, 30
induction charging, 29-30, 39
inductive coupling, 7, 66
inspection procedures, ESD-sensitive devices, 170-171
insulators, 23-26, 30
interference (*see* electromagnetic/radio frequency interference)
International Electrotechnical Commission (IEC), 78
ionizers/ion imbalance, 1, 27-28, 68, 94, 104-109
    air ionizers, 104-108
    electric-powered, 105
    hazards associated with, 107-108
    hot vs. cold bar, 106-107
    ionizing air guns, 109
    nuclear, 108-109
    ozone hazard and, 105-106, 108
    static comb, 104
ITT Research Institute, 11

J

Jet Propulsion Laboratory, 11
JFETs, 7, 62
junction burnout, 61
junction current-voltage trace, 64

K

KeyTek Instrument Corp., 151, 155-158

Index

L

labels, anti-ESD warning, 113, 166, 167
lightning, 10, 12, 13, 14, 44, 66
linear circuit ICs, 57, 66
lockup, 61, 66
lubricity, 22, 79, 81

M

magnetic fields (H-field), 152
metal-oxide varistors (MOV), 4, 74, 75
metallization melt, 63-64, 67
metalloplastics, 26
microwave transistors, 7
MIL-STD-883, 53-54, 60, 110, 150-155, 168
mobile charges, 22, 30
monitors, electrostatic, 126, 140-141
monolithic microcircuits, 57
monolithic multilayer ceramic capacitors (MLC), 48-49
Monroe equipment, 130-133, 136-137, 140
MOS transistors, 6, 7, 8, 57, 62, 63, 69, 70, 167-169
MOSFETs, 7, 16, 63, 65
multiple-device tester/simulator, 158

N

National Electrical Manufacturers Association (NEMA), 153
National Fire Protection Association (NFPA), 27
neutrons, 16-17
NMOS, 7, 57, 66
nuclear electromagnetic pulses (NEMP), 5, 12, 13, 14
nuclear fieldmeter, 129
nuclear static eliminator (*see also* ionizers), 108-109

O

op amps, 7, 57, 61, 67
open circuits, 63
overcurrent, 6
oxide punchthrough, 62-63, 69
ozone, ionizer hazard, 105-106, 108

P

packaging (*see* containers and packaging and)
parallel plate capacitance/capacitors, 46
parasitic capacitors, 49-50
personal ESD protection, 114-125
    ESD-protective clothing, 121-124
    field service kits, 124-125
    gloves and finger cots, 124
    shoe straps, 120-121
    wrist straps, 114-120
piezoelectric crystals, 7, 57
PMOS, 7, 57, 66
point charges, 19
polarity, 39, 47
portable simulators, 155-157
potential difference, 40, 41, 44
potential gradient, 130
power surges, 66
printed circuit boards, 7, 8, 27, 57, 77
    coupling prevention, 75
    handling guidelines, 169-170
    inspection procedures, 170-171
    metal-oxide varistors (MOV), 74, 75
    protective devices for, 72-76
    protective networks for, 72-73
    testing procedures, 171
    tote boxes/storage cases for, 88-90
    transient suppressors on, 72-74
    transient voltage suppressors (TVS), 74
product design, ESD vs., 174-175
protective bags, 83-85
protective devices, 4, 8, 69-76, 149, 150
    antenna effect vs., 149
    capacitors in, 69
    current-limiting, 69
    diodes in, 69
    filters, 75
    gate density and, 70
    metal-oxide varistors (MOV), 74, 75
    protective networks, 69, 70, 72-73
    resistors in, 69
    transient suppressors, 72-74
    transient voltage suppressors (TVS), 74
    voltage-clamping, 69, 70, 73, 75, 149
protective networks, 4, 8, 53, 69, 70, 72-73
protons, 16-17
punchthrough, 62-63, 69

R

radiated coupling, 6
radio frequency interference (RFI), 13

Index

RC networks, 53
relative humidity, 1, 9, 11, 15, 20-24, 26, 28, 55, 68, 69, 79, 81, 82, 93-94, 145-146
relays, 7
Reliability Anaysis Center, 11
repairs, cost of, 2-3, 8
repulsion, electrostatics, 17-19, 24, 32, 37-40, 104
resistance/resistivity
  sheet, 25
  surface resistivity, 25, 26, 78, 79, 126, 142, 143
  volume resistivity, 24, 25, 78, 142, 143
resistors, 7, 57, 69
rf devices, 66
rf radiation, 151
Richmond Corp., 11

S

scanning electron microscope (SEM), 9, 65
Schottky diodes, 7, 69
SCRs, 7
seating/cushions, ESD-protective, 104
secondary breakdown, 6, 61-62
semiconductors
  damage detection, 2, 6, 8-10, 60-66
  design of vs. ESD, 2-5, 7-9, 11, 56-61, 66, 69-76, 149, 150
  handling guidelines, 167-169
  hard failure, 60-64
  immediate vs. delayed failure, 60
  inspection procedures, 170-171
  lockup, 61
  permanent vs. temporary failure, 60
  protective devices for, 4, 8, 69-72, 149, 150
  soft failure, 60, 61, 65-67
  susceptibility classifications, 57-61
  testing procedures, 171
  upset, 61
sheet resistance, 25
shielded bag test systems, 126, 146-147
shielding, 76, 81
shoe straps, 4, 102, 120-121, 147, 167
short circuits, 63
shorting plugs, 77
shunt bars, 91-92
signal conditioners, 57

signs, anti-ESD warning, 113, 166, 167
simulators, 149-158
  bench-type, 154
  commercial, 153-158
  hand-held and portable, 155-157
  hand/metal ESDs, 151
  human body ESD, 50-55, 150, 152, 153
  multiple-device tester, 158
  standards for, 150-153
Society of Automotive Engineers (SAE), 153
soft failure, 60, 61, 65-67
soft grounds, 28, 29, 95
software, ESD damage prevention, 76
standards, 56, 78, 152-153, 176-177
  antistatic materials, 78-83, 95, 96
  conductive materials, 78-83, 95, 96
  semiconductor design/ESD susceptibility, 56
  simulators and testing, 150-153
  static-dissipative material, 78-83, 95, 96
static comb ionizers, 104
static decay meters, 126, 143-144
static electricity, 15
Static Inc., 84-85, 96
static-dissipative material (*see also* containers and packaging), 26, 78-83, 95, 96
storage cases, ESD-protective, 88-90
suppliers, 178-198
surface acoustic wave (SAW) devices, 7, 57
surface charge density, 43-44
surface resistivity, 25, 26, 78, 79, 126, 142, 143
surface voltage, 128
surface/volume resistivity probes, 126, 142, 143
systems, protective devices for, 75-76

T

test equipment, 126-148
  ac-carrier fieldmeter, 129
  chopper-stabilized fieldmeter, 129
  electrometer, 129
  fieldmeters, 126
  ground strap testers, 126, 147-148
  hand-held field- and voltmeters, 133-137
  humidity test chambers, 126, 145-146

Index

monitors, 126, 140-141
nuclear fieldmeter, 129
portable survey and audit, 127
shielded bag test systems, 126, 146-147
static decay meters, 126, 143-144
surface/volume resistivity probes, 126, 142, 143
voltmeters, 126-140
test equipment/tools, grounding, 99
testing procedures, ESD-sensitive devices, 171
thermal secondary breakdown, 61-62
thin-film resistors, 7
topical antistats, 111
tote boxes, ESD-protective, 88-90
training, 159, 173-174
transient suppressors, 72-74
transient voltage suppressors (TVS), 4, 74
transistor-transistor logic (TTL), 7, 66, 69
transistors, 4
  bipolar, 6, 7, 57, 61, 62, 69
  breakdown voltages, 69
  gallium arsenide, 57

microwave, 7
MOS, 6, 7, 8, 57, 62, 63, 69, 70, 167-169
MOSFETs, 7, 16, 63, 65
triboelectric charging, 20-24, 77, 79, 126

U

Underwriter's Laboratories (UL), 153
unintentional capacitors, 49-50
upset, 61, 65-67

V

vacuum tubes, 4
Van de Graaf generator, 44, 50-52
very large scale (VLSI) circuits, 2, 7
voltage, 41, 44, 128
voltage-clamping devices, 69, 70, 73, 75, 149
voltmeters, 126-140
volume resistivity, 24, 25, 78, 142, 143

W

wrist straps, 1, 4, 9, 29, 68, 83, 95, 114-120, 147, 161, 166, 167